S0-ARK-966

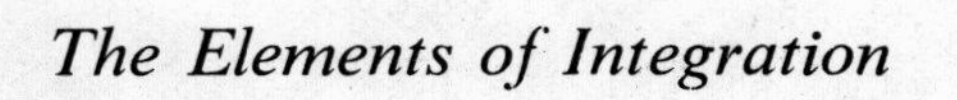

By the Same Author

The Elements of Real Analysis

The Elements of Integration

ROBERT G. BARTLE

Professor of Mathematics
University of Illinois

JOHN WILEY & SONS, New York • Chichester • Brisbane • Toronto

Copyright © 1966 by John Wiley & Sons, Inc.
All rights reserved.
Reproduction or translation of any part of this work beyond that permitted by Sections 107 or 108 of the 1976 United States Copyright Act without the permission of the copyright owner is unlawful. Requests for permission or further information should be addressed to the Permissions Department, John Wiley & Sons, Inc.

ISBN 0 471 05457 7

Library of Congress Catalog Card Number: 66–21048
Printed in the United States of America

10 9 8

Preface

This book was written to present to a reader having only a modest mathematical background the chief results in the modern theory of integration which was initiated by Lebesgue in 1902. Lebesgue's integral has now become one of the cornerstones of mathematical analysis. This book developed from my lectures in a course at the University of Illinois and should be accessible to advanced undergraduate and beginning graduate students; its prerequisites are an understanding of the elementary theory of real analysis and the ability to comprehend "$\varepsilon - \delta$ arguments." Although it is likely that a reader would have some familiarity with the Riemann integral, I do not presuppose that he has mastered its theoretical details, for the presentation given here does not depend on the Riemann integral. A solid course in "advanced calculus," or familiarity with the first third of my book, *The Elements of Real Analysis*, should provide adequate background for reading this book.

It has been my experience, both as student and teacher, that most students have difficulty in seeing the subject as a whole, and that surprisingly many have troubles with some of its major parts. I suspect that this is partly attributable to the different approaches to integration theory, but I think that it may also be due to the character of the current texts. Most authors who treat abstract measure spaces start with fairly extensive and detailed discussions of measure theory; only later do they turn to integration. This tends to give the impression that an elaborate theory of measure is required for an understanding of integration. I feel that this is no more true than that a detailed study of set theory is required for an understanding of topology.

Other authors prefer to get to the integral and its properties quickly. Such books often start with some type of "elementary integral" and extend it to a larger class of functions, after which they obtain whatever measure theory they wish. Their point of departure varies widely. It may be the Riemann integral on the continuous functions on an interval or a rectangle; it may be a linear functional on a collection of continuous functions; it may be an abstractly defined integral on a class of functions. Usually, these treatments inject topological notions at an early stage where, in my judgment, topology is neither needed nor desired.

Although I also wish to develop the integral as soon as it can be conveniently done, I prefer to discuss abstract measure spaces. I regard the convergence theorems as the *raison d'être* for the theory and consider set theory, measure theory, and topology to be largely irrelevant—they cannot be completely disregarded, but they should not be given undue prominence at the outset, for they only complicate the situation. However, once the initial steps have been taken and the integral has been established, the reader should try to connect the integral with other parts of mathematics.

Integration theory is much like point set topology: it is a basic subject, but it is not an end in itself. My purpose has been to strike directly toward the main results; I have not attempted to follow all the avenues which have been opened. Thus, a reader who completes this book is not through; instead, I hope that he will delve into the many questions that I have purposely laid aside. I put them aside because I feel that these questions are not truly relevant to an introductory development of the basic ideas of integration; this does not mean that they are trivial or uninteresting. There is much that is yet to be done for the reader, but if this book helps speed him on his way, it has accomplished its purpose.

Since this is intended as an introduction, I shall deal with real-valued functions and with countably additive measures. Until recently it was thought that countable additivity was a necessary ingredient of any "decent" theory of integration, but I believe this has been fully exploded by Chapter III of the treatise of Dunford and Schwartz [5]. Although various theories of vector-valued integration are available, I do not touch upon this subject, but refer the reader to References [2] and [5], and the papers cited there. For treatments of the "abstract" theory of integration, I recommend References [3, 5, 7, 10, 12, 13]. For the "elementary integral" approach, I suggest that the reader first consult [14] for orientation, and then sample References [4, 6, 8, 9, 11, 15]. In [16, 18] both approaches are developed and related to each other.

In writing this book, I have benefited greatly from my teachers, colleagues, and students. Most of what I present is derived from what I learned from some member of one of these three groups of people. I am particularly indebted to my colleague A. L. Peressini, who read an earlier draft of this manuscript. Professor George Orland made several suggestions, enabling me to strengthen certain theorems and to correct other proofs; Professor N. T. Hamilton proposed an example (now Exercise 10.S) one Saturday afternoon over coffee. Mrs. Carolyn Bloemker ably typed the final version of the manuscript, and the galley proofs were checked by Mr. Charles W. Mullins, who caught a number of errors. To all of these people, I am deeply indebted. In addition, I greatly appreciate the cooperation of the staff at John Wiley and Sons for their help and consideration.

ROBERT G. BARTLE

Champaign-Urbana, Illinois

April 1, 1966

Contents

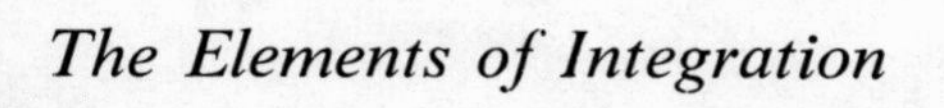

The Elements of Integration

CHAPTER 1

Introduction

The theory of integration has its ancient and honorable roots in the "method of exhaustion" that was invented by Eudoxos and greatly developed by Archimedes for the purpose of calculating the areas and volumes of geometric figures. The later work of Newton and Leibniz enabled this method to grow into a systematic tool for such calculations.

As this theory developed, it has become less concerned with applications to geometry and elementary mechanics, for which it is entirely adequate, and more concerned with purely analytic questions, for which the classical theory of integration is not always sufficient. Thus a present-day mathematician is apt to be interested in the convergence of orthogonal expansions, or in applications to differential equations or probability. For him the classical theory of integration which culminated in the Riemann integral has been largely replaced by the theory which has grown from the pioneering work of Henri Lebesgue at the beginning of this century. The reason for this is very simple: the powerful convergence theorems associated with the Lebesgue theory of integration lead to more general, more complete, and more elegant results than the Riemann integral admits.

Lebesgue's definition of the integral enlarges the collection of functions for which the integral is defined. Although this enlargement is useful in itself, its main virtue is that the theorems relating to the interchange of the limit and the integral are valid under less stringent assumptions than are required for the Riemann integral. Since one frequently needs to make such interchanges, the Lebesgue integral is more convenient to deal with than the Riemann integral. To exemplify these remarks, let the sequence (f_n) of functions be defined for $x > 0$ by $f_n(x) = e^{-nx}/\sqrt{x}$. It is readily seen that the (improper) Riemann integrals

$$I_n = \int_0^{+\infty} \frac{e^{-nx}}{\sqrt{x}} \, dx$$

exist and that $\lim_{n\to\infty} f_n(x) = 0$ for all $x > 0$. However, since $\lim_{x\to 0} f_n(x) = +\infty$ for each n, the convergence of the sequence is certainly not uniform for $x > 0$. Although it is hoped that the reader can supply the estimates required to show that $\lim I_n = 0$, we prefer to obtain this conclusion as an immediate consequence of the Lebesgue Dominated Convergence Theorem which will be proved later. As another example, consider the function F defined for $t > 0$ by the (improper) Riemann integral

$$F(t) = \int_0^{+\infty} x^2 e^{-tx} \, dx.$$

With a little effort one can show that F is continuous and that its derivative exists and is given by

$$F'(t) = -\int_0^{+\infty} x^3 e^{-tx} \, dx,$$

which is obtained by differentiating under the integral sign. Once again, this inference follows easily from the Lebesgue Dominated Convergence Theorem.

At the risk of oversimplification, we shall try to indicate the crucial difference between the Riemann and the Lebesgue definitions of the integral. Recall that an **interval** in the set $\boldsymbol{R}$ of real numbers is a set which has one of the following four forms:

$$[a, b] = \{x \in \boldsymbol{R} : a \leqslant x \leqslant b\}, \qquad (a, b) = \{x \in \boldsymbol{R} : a < x < b\},$$
$$[a, b) = \{x \in \boldsymbol{R} : a \leqslant x < b\}, \qquad (a, b] = \{x \in \boldsymbol{R} : a < x \leqslant b\}.$$

In each of these cases we refer to a and b as the **endpoints** and prescribe

$b - a$ as the **length** of the interval. Recall further that if E is a set, then the **characteristic function** of E is the function χ_E defined by

$$\begin{aligned} \chi_E(x) &= 1, \qquad \text{if } x \in E, \\ &= 0, \qquad \text{if } x \notin E. \end{aligned}$$

A **step function** is a function φ which is a finite linear combination of characteristic functions of intervals; thus

$$\varphi = \sum_{j=1}^{n} c_j \chi_{E_j}.$$

If the endpoints of the interval E_j are a_j, b_j, we define the **integral** of φ to be

$$\int \varphi = \sum_{j=1}^{n} c_j(b_j - a_j).$$

If f is a bounded function defined on an interval $[a, b]$ and if f is not too discontinuous, then the **Riemann integral** of f is defined to be the limit (in an appropriate sense) of the integrals of step functions which approximate f. In particular, the **lower Riemann integral** of f may be defined to be the supremum of the integrals of all step functions φ such that $\varphi(x) \leqslant f(x)$ for all x in $[a, b]$, and $\varphi(x) = 0$ for x not in $[a, b]$.

The Lebesgue integral can be obtained by a similar process, except that the collection of step functions is replaced by a larger class of functions. In somewhat more detail, the notion of length is generalized to a suitable collection $\boldsymbol{X}$ of subsets of $\boldsymbol{R}$. Once this is done, the step functions are replaced by **simple functions**, which are finite linear combinations of characteristic functions of sets belonging to $\boldsymbol{X}$. If

$$\varphi = \sum_{j=1}^{n} c_j \chi_{E_j}$$

is such a simple function and if $\mu(E)$ denotes the "measure" or "generalized length" of the set E in $\boldsymbol{X}$, we define the integral of φ to be

$$\int \varphi = \sum_{j=1}^{n} c_j \mu(E_j).$$

If f is a nonnegative function defined on $\boldsymbol{R}$ which is suitably restricted, we shall define the **(Lebesgue) integral** of f to be the supremum of the integrals of all simple functions φ such that $\varphi(x) \leqslant f(x)$ for all x in $\boldsymbol{R}$. The integral can then be extended to certain functions that take both signs.

Although the generalization of the notion of length to certain sets in $\boldsymbol{R}$ which are not necessarily intervals has great interest, it was observed in 1915 by Maurice Fréchet that the convergence properties of the Lebesgue integral are valid in considerable generality. Indeed, let X be any set in which there is a collection $\boldsymbol{X}$ of subsets containing the empty set $\emptyset$ and X and closed under complementation and countable unions. Suppose that there is a nonnegative measure function μ defined on $\boldsymbol{X}$ such that $\mu(\emptyset) = 0$ and which is **countably additive** in the sense that

$$\mu\left(\bigcup_{j=1}^{\infty} E_j\right) = \sum_{j=1}^{\infty} \mu(E_j)$$

for each sequence (E_j) of sets in $\boldsymbol{X}$ which are mutually disjoint. In this case an integral can be defined for a suitable class of real-valued functions on X, and this integral possesses strong convergence properties.

As we have stressed, we are particularly interested in these convergence theorems. Therefore we wish to advance directly toward them in this abstract setting, since it is more general and, we believe, conceptually simpler than the special cases of integration on the line or in $\boldsymbol{R}^n$. However, it does require that the reader temporarily accept the fact that interesting special cases are subsumed by the general theory. Specifically, it requires that he accept the assertion that there exists a countably additive measure function that extends the notion of the length of an interval. The proof of this assertion is in Chapter 9 and can be read after completing Chapter 3 by those for whom the suspense is too great.

In this introductory chapter we have attempted to provide motivation and to set the stage for the detailed discussion which follows. Some of our remarks here have been a bit vague and none of them has been proved. These defects will be remedied. However, since we shall have occasion to refer to the system of extended real numbers, we now append a brief description of this system.

In integration theory it is frequently convenient to adjoin the two symbols $-\infty$, $+\infty$ to the real number system $\boldsymbol{R}$. (It is stressed that these symbols are not real numbers.) We also introduce the convention that $-\infty < x < +\infty$ for any $x \in \boldsymbol{R}$. The collection $\bar{\boldsymbol{R}}$ consisting of the set $\boldsymbol{R} \cup \{-\infty, +\infty\}$ is called the **extended real number system.**

One reason we wish to consider $\bar{\boldsymbol{R}}$ is that it is convenient to say that the length of the real line is equal to $+\infty$. Another reason is that we will frequently be taking the supremum (= least upper bound) of a set of real numbers. We know that a nonempty set A of real numbers which has an upper bound also has a supremum (in $\boldsymbol{R}$). If we define the supremum of a nonempty set which does not have an upper bound to be $+\infty$, then every nonempty subset of $\boldsymbol{R}$ (or $\bar{\boldsymbol{R}}$) has a unique supremum in $\bar{\boldsymbol{R}}$. Similarly, every nonempty subset of $\boldsymbol{R}$ (or $\bar{\boldsymbol{R}}$) has a unique infimum (= greatest lower bound) in $\bar{\boldsymbol{R}}$. (Some authors introduce the conventions that $\inf \emptyset = +\infty$, $\sup \emptyset = -\infty$, but we shall not employ them.)

If (x_n) is a sequence of extended real numbers, we define the **limit superior** and the **limit inferior** of this sequence by

$$\limsup x_n = \inf_m \Big(\sup_{n \geqslant m} x_n \Big),$$

$$\liminf x_n = \sup_m \Big(\inf_{n \geqslant m} x_n \Big).$$

If the limit inferior and the limit superior are equal, then their value is called the **limit** of the sequence. It is clear that this agrees with the conventional definition when the sequence and the limit belong to $\boldsymbol{R}$.

Finally, we introduce the following algebraic operations between the symbols $\pm\infty$ and elements $x \in \boldsymbol{R}$:

$$\begin{aligned}(\pm\infty) + (\pm\infty) = x + (\pm\infty) = (\pm\infty) + x &= \pm\infty, \\ (\pm\infty)(\pm\infty) = +\infty, \ (\pm\infty)(\mp\infty) &= -\infty, \\ x(\pm\infty) = (\pm\infty)x = \pm\infty \quad & \text{if } x > 0, \\ = 0 \quad & \text{if } x = 0, \\ = \mp\infty \quad & \text{if } x < 0.\end{aligned}$$

It should be noticed that we do not define $(+\infty) + (-\infty)$ or $(-\infty) + (+\infty)$, nor do we define quotients when the denominator is $\pm\infty$.

CHAPTER 2

Measurable Functions

In developing the Lebesgue integral we shall be concerned with classes of real-valued functions defined on a set X. In various applications the set X may be the unit interval $\boldsymbol{I} = [0, 1]$ consisting of all real numbers x satisfying $0 \leqslant x \leqslant 1$; it may be the set $\boldsymbol{N} = \{1, 2, 3, \ldots\}$ of natural numbers; it may be the entire real line $\boldsymbol{R}$; it may be all of the plane; or it may be some other set. Since the development of the integral does not depend on the character of the underlying space X, we shall make no assumptions about its specific nature.

Given the set X, we single out a family $\boldsymbol{X}$ of subsets of X which are "well-behaved" in a certain technical sense. To be precise, we shall assume that this family contains the empty set $\emptyset$ and the entire set X, and that $\boldsymbol{X}$ is closed under complementation and countable unions.

2.1 Definition. A family $\boldsymbol{X}$ of subsets of a set X is said to be a **σ-algebra** (or a **σ-field**) in case:

(i) $\emptyset$, X belong to $\boldsymbol{X}$.

(ii) If A belongs to $\boldsymbol{X}$, then the complement $\mathscr{C}(A) = X \setminus A$ belongs to $\boldsymbol{X}$.

(iii) If (A_n) is a sequence of sets in $\boldsymbol{X}$, then the union $\bigcup_{n=1}^{\infty} A_n$ belongs to $\boldsymbol{X}$.

An ordered pair $(X, \boldsymbol{X})$ consisting of a set X and a σ-algebra $\boldsymbol{X}$ of subsets of X is called a **measurable space**. Any set in $\boldsymbol{X}$ is called an

$\boldsymbol{X}$-measurable set, but when the σ-algebra $\boldsymbol{X}$ is fixed (as is generally the case), the set will usually be said to be **measurable**.

The reader will recall the rules of De Morgan:

$$\text{(2.1)} \qquad \mathscr{C}\Big(\bigcup_{\alpha} A_\alpha\Big) = \bigcap_{\alpha} \mathscr{C}(A_\alpha), \qquad \mathscr{C}\Big(\bigcap_{\alpha} A_\alpha\Big) = \bigcup_{\alpha} \mathscr{C}(A_\alpha).$$

It follows from these that the intersection of a sequence of sets in $\boldsymbol{X}$ also belongs to $\boldsymbol{X}$.

We shall now give some examples of σ-algebras of subsets.

2.2 Examples. (a) Let X be any set and let $\boldsymbol{X}$ be the family of all subsets of X.

(b) Let $\boldsymbol{X}$ be the family consisting of precisely two subsets of X, namely $\emptyset$ and X.

(c) Let $X = \{1, 2, 3, \ldots\}$ be the set $\boldsymbol{N}$ of natural numbers and let $\boldsymbol{X}$ consist of the subsets

$$\emptyset, \quad \{1, 3, 5, \ldots\}, \qquad \{2, 4, 6, \ldots\}, \qquad X.$$

(d) Let X be an uncountable set and $\boldsymbol{X}$ be the collection of subsets which are either countable or have countable complements.

(e) If $\boldsymbol{X}_1$ and $\boldsymbol{X}_2$ are σ-algebras of subsets of X, let $\boldsymbol{X}_3$ be the intersection of $\boldsymbol{X}_1$ and $\boldsymbol{X}_2$; that is, $\boldsymbol{X}_3$ consists of all subsets of X which belong to both $\boldsymbol{X}_1$ and $\boldsymbol{X}_2$. It is readily checked that $\boldsymbol{X}_3$ is a σ-algebra.

(f) Let $\boldsymbol{A}$ be a nonempty collection of subsets of X. We observe that there is a smallest σ-algebra of subsets of X containing $\boldsymbol{A}$. To see this, observe that the family of all subsets of X is a σ-algebra containing $\boldsymbol{A}$ and the intersection of all the σ-algebras containing $\boldsymbol{A}$ is also a σ-algebra containing $\boldsymbol{A}$. This smallest σ-algebra is sometimes called the **σ-algebra generated by $\boldsymbol{A}$**.

(g) Let X be the set $\boldsymbol{R}$ of real numbers. The **Borel algebra** is the σ-algebra $\boldsymbol{B}$ generated by all open intervals (a, b) in $\boldsymbol{R}$. Observe that the Borel algebra $\boldsymbol{B}$ is also the σ-algebra generated by all closed intervals $[a, b]$ in $\boldsymbol{R}$. Any set in $\boldsymbol{B}$ is called a **Borel set**.

(h) Let X be the set $\bar{\boldsymbol{R}}$ of extended real numbers. If E is a Borel subset of $\boldsymbol{R}$, let

$$\text{(2.2)} \quad E_1 = E \cup \{-\infty\}, \qquad E_2 = E \cup \{+\infty\}, \qquad E_3 = E \cup \{-\infty, +\infty\},$$

and let $\bar{\boldsymbol{B}}$ be the collection of all sets E, E_1, E_2, E_3 as E varies over $\boldsymbol{B}$. It is readily seen that $\bar{\boldsymbol{B}}$ is a σ-algebra and it will be called the **extended Borel algebra.**

In the following, we shall consider a fixed measurable space $(X, \boldsymbol{X})$.

2.3 DEFINITION. A function f on X to $\boldsymbol{R}$ is said to be ***X*-measurable** (or simply **measurable**) if for every real number α the set

$$\{x \in X : f(x) > \alpha\} \tag{2.3}$$

belongs to $\boldsymbol{X}$.

The next lemma shows that we could have modified the form of the sets in defining measurability.

2.4 LEMMA. *The following statements are equivalent for a function f on X to $\boldsymbol{R}$:*

(a) *For every $\alpha \in \boldsymbol{R}$, the set $A_\alpha = \{x \in X : f(x) > \alpha\}$ belongs to $\boldsymbol{X}$.*
(b) *For every $\alpha \in \boldsymbol{R}$, the set $B_\alpha = \{x \in X : f(x) \leqslant \alpha\}$ belongs to $\boldsymbol{X}$.*
(c) *For every $\alpha \in \boldsymbol{R}$, the set $C_\alpha = \{x \in X : f(x) \geqslant \alpha\}$ belongs to $\boldsymbol{X}$.*
(d) *For every $\alpha \in \boldsymbol{R}$, the set $D_\alpha = \{x \in X : f(x) < \alpha\}$ belongs to $\boldsymbol{X}$.*

PROOF. Since B_α and A_α are complements of each other, statement (a) is equivalent to statement (b). Similarly, statements (c) and (d) are equivalent. If (a) holds, then $A_{\alpha - 1/n}$ belongs to $\boldsymbol{X}$ for each n and since

$$C_\alpha = \bigcap_{n=1}^{\infty} A_{\alpha - 1/n},$$

it follows that $C_\alpha \in \boldsymbol{X}$. Hence (a) implies (c). Since

$$A_\alpha = \bigcup_{n=1}^{\infty} C_{\alpha + 1/n},$$

it follows that (c) implies (a). Q.E.D.

2.5 EXAMPLES. (a) Any constant function is measurable. For, if $f(x) = c$ for all $x \in X$ and if $\alpha \geqslant c$, then

$$\{x \in X : f(x) > \alpha\} = \emptyset,$$

whereas if $\alpha < c$, then

$$\{x \in X : f(x) > \alpha\} = X.$$

(b) If $E \in \boldsymbol{X}$, then the **characteristic function** χ_E, defined by

$$\begin{aligned} \chi_E(x) &= 1, \quad x \in E, \\ &= 0, \quad x \notin E, \end{aligned}$$

is measurable. In fact, $\{x \in X : \chi_E(x) > \alpha\}$ is either X, E, or $\emptyset$.

(c) If X is the set $\boldsymbol{R}$ of real numbers, and $\boldsymbol{X}$ is the Borel algebra $\boldsymbol{B}$, then any continuous function f on $\boldsymbol{R}$ to $\boldsymbol{R}$ is Borel measurable (that is, $\boldsymbol{B}$-measurable). In fact, if f is continuous, then $\{x \in \boldsymbol{R} : f(x) > \alpha\}$ is an open set in $\boldsymbol{R}$ and hence is the union of a sequence of open intervals. Therefore, it belongs to $\boldsymbol{B}$.

(d) If $X = \boldsymbol{R}$ and $\boldsymbol{X} = \boldsymbol{B}$, then any monotone function is Borel measurable. For, suppose that f is monotone increasing in the sense that $x \leqslant x'$ implies $f(x) \leqslant f(x')$. Then $\{x \in \boldsymbol{R} : f(x) > \alpha\}$ consists of a half-line which is either of the form $\{x \in \boldsymbol{R} : x > a\}$ or the form $\{x \in \boldsymbol{R} : x \geqslant a\}$. (Show that both cases can occur.)

Certain simple algebraic combinations of measurable functions are measurable, as we shall now show.

2.6 LEMMA. *Let f and g be measurable real-valued functions and let c be a real number. Then the functions*

$$cf, \quad f^2, \quad f + g, \quad fg, \quad |f|,$$

are also measurable.

PROOF. (a) If $c = 0$, the statement is trivial. If $c > 0$, then

$$\{x \in X : cf(x) > \alpha\} = \{x \in X : f(x) > \alpha/c\} \in \boldsymbol{X}.$$

The case $c < 0$ is handled similarly.

(b) If $\alpha < 0$, then $\{x \in X : (f(x))^2 > \alpha\} = X$; if $\alpha \geqslant 0$, then

$$\begin{aligned} \{x \in X : &(f(x))^2 > \alpha\} \\ &= \{x \in X : f(x) > \sqrt{\alpha}\} \cup \{x \in X : f(x) < -\sqrt{\alpha}\}. \end{aligned}$$

(c) By hypothesis, if r is a rational number, then

$$S_r = \{x \in X : f(x) > r\} \cap \{x \in X : g(x) > \alpha - r\}$$

belongs to $\boldsymbol{X}$. Since it is readily seen that

$$\{x \in X : (f + g)(x) > \alpha\} = \bigcup \{S_r : r \text{ rational}\},$$

it follows that $f + g$ is measurable.

(d) Since $fg = \frac{1}{4}[(f+g)^2 - (f-g)^2]$, it follows from parts (a), (b), and (c) that fg is measurable.

(e) If $\alpha < 0$, then $\{x \in X : |f(x)| > \alpha\} = X$, whereas if $\alpha \geqslant 0$, then

$$\{x \in X : |f(x)| > \alpha\} = \{x \in X : f(x) > \alpha\} \cup \{x \in X : f(x) < -\alpha\}.$$

Thus the function $|f|$ is measurable. Q.E.D.

If f is any function on X to $\boldsymbol{R}$, let f^+ and f^- be the nonnegative functions defined on X by

$$(2.4) \qquad f^+(x) = \sup\{f(x), 0\}, \quad f^-(x) = \sup\{-f(x), 0\}.$$

The function f^+ is called the **positive part** of f and f^- is called the **negative part** of f. It is clear that

$$(2.5) \qquad f = f^+ - f^- \quad \text{and} \quad |f| = f^+ + f^-$$

and it follows from these identities that

$$(2.6) \qquad f^+ = \tfrac{1}{2}(|f| + f), \quad f^- = \tfrac{1}{2}(|f| - f).$$

In view of the preceding lemma we infer that f is measurable if and only if f^+ and f^- are measurable.

The preceding discussion pertained to real-valued functions defined on a measurable space. However, in dealing with sequences of measurable functions we often wish to form suprema, limits, etc., and it is technically convenient to allow the extended real numbers $-\infty$, $+\infty$ to be taken as values. Hence we wish to define measurability for extended real-valued functions and we do this exactly as in Definition 2.3.

2.7 Definition. An extended real-valued function on X is **$\boldsymbol{X}$-measurable** in case the set $\{x \in X : f(x) > \alpha\}$ belongs to $\boldsymbol{X}$ for each real number α. The collection of all extended real-valued $\boldsymbol{X}$-measurable functions on X is denoted by $M(X, \boldsymbol{X})$.

Observe that if $f \in M(X, \boldsymbol{X})$, then

$$\{x \in X : f(x) = +\infty\} = \bigcap_{n=1}^{\infty} \{x \in X : f(x) > n\},$$

$$\{x \in X : f(x) = -\infty\} = \mathscr{C}\Big[\bigcup_{n=1}^{\infty} \{x \in X : f(x) > -n\}\Big],$$

so that both of these sets belong to $\boldsymbol{X}$.

The following lemma is often useful in treating extended real-valued functions.

2.8 Lemma. *An extended real-valued function f is measurable if and only if the sets*

$$A = \{x \in X : f(x) = +\infty\}, \quad B = \{x \in X : f(x) = -\infty\}$$

belong to $\boldsymbol{X}$ and the real-valued function f_1 defined by

$$\begin{aligned} f_1(x) &= f(x), \qquad && \text{if } x \notin A \cup B, \\ &= 0, && \text{if } x \in A \cup B, \end{aligned}$$

is measurable.

PROOF. If f is in $M(X, \boldsymbol{X})$, it has already been noted that A and B belong to $\boldsymbol{X}$. Let $\alpha \in \boldsymbol{R}$ and $\alpha \geqslant 0$, then

$$\{x \in X : f_1(x) > \alpha\} = \{x \in X : f(x) > \alpha\} \setminus A.$$

If $\alpha < 0$, then

$$\{x \in X : f_1(x) > \alpha\} = \{x \in X : f(x) > \alpha\} \cup B.$$

Hence f_1 is measurable.

Conversely, if $A, B \in \boldsymbol{X}$ and f_1 is measurable, then

$$\{x \in X : f(x) > \alpha\} = \{x \in X : f_1(x) > \alpha\} \cup A$$

when $\alpha \geqslant 0$, and

$$\{x \in X : f(x) > \alpha\} = \{x \in X : f_1(x) > \alpha\} \setminus B$$

when $\alpha < 0$. Therefore f is measurable. Q.E.D.

It is a consequence of Lemmas 2.6 and 2.8 that if f is in $M(X, \boldsymbol{X})$, then the functions

$$cf, \quad f^2, \quad |f|, \quad f^+, \quad f^-$$

also belong to $M(X, \boldsymbol{X})$.

The only comment that need be made is that we adopt the convention that $0(\pm\infty) = 0$ so that cf vanishes identically when $c = 0$. If f and g belong to $M(X, \boldsymbol{X})$, then the sum $f + g$ is not well-defined by the formula $(f + g)(x) = f(x) + g(x)$ on the sets

$$\begin{aligned} E_1 &= \{x \in X : f(x) = -\infty \quad \text{and} \quad g(x) = +\infty\}, \\ E_2 &= \{x \in X : f(x) = +\infty \quad \text{and} \quad g(x) = -\infty\}, \end{aligned}$$

both of which belong to $\boldsymbol{X}$. However, if we define $f + g$ to be zero on $E_1 \cup E_2$, the resulting function on X is measurable. We shall return to the measurability of the product fg after the next result.

2.9 LEMMA. *Let (f_n) be a sequence in $M(X, \boldsymbol{X})$ and define the functions*

$$f(x) = \inf f_n(x), \quad F(x) = \sup f_n(x),$$
$$f^*(x) = \liminf f_n(x), \quad F^*(x) = \limsup f_n(x).$$

Then f, F, f^*, and F^* belong to $M(X, \boldsymbol{X})$.

PROOF. Observe that

$$\{x \in X : f(x) \geqslant \alpha\} = \bigcap_{n=1}^{\infty} \{x \in X : f_n(x) \geqslant \alpha\},$$

$$\{x \in X : F(x) > \alpha\} = \bigcup_{n=1}^{\infty} \{x \in X : f_n(x) > \alpha\},$$

so that f and F are measurable when all the f_n are. Since

$$f^*(x) = \sup_{n \geqslant 1} \left\{ \inf_{m \geqslant n} f_m(x) \right\},$$

$$F^*(x) = \inf_{n \geqslant 1} \left\{ \sup_{m \geqslant n} f_m(x) \right\},$$

the measurability of f^* and F^* is also established. Q.E.D.

2.10 COROLLARY. *If (f_n) is a sequence in $M(X, \boldsymbol{X})$ which converges to f on X, then f is in $M(X, \boldsymbol{X})$.*

PROOF. In this case $f(x) = \lim f_n(x) = \liminf f_n(x)$. Q.E.D.

We now return to the measurability of the product $f\,g$ when f, g belong to $M(X, \boldsymbol{X})$. If $n \in \boldsymbol{N}$, let f_n be the "truncation of f" defined by

$$\begin{aligned} f_n(x) &= f(x), && \text{if } |f(x)| \leqslant n, \\ &= n, && \text{if } f(x) > n, \\ &= -n, && \text{if } f(x) < -n. \end{aligned}$$

Let g_m be defined similarly. It is readily seen that f_n and g_m are measurable (see Exercise 2.K). It follows from Lemma 2.6 that the product $f_n\, g_m$ is measurable. Since

$$f(x)\, g_m(x) = \lim_n f_n(x)\, g_m(x), \quad x \in X,$$

it follows from Corollary 2.10 that fg_m belongs to $M(X, \mathbf{X})$. Since

$$(fg)(x) = f(x)\,g(x) = \lim_m f(x)\,g_m(x), \quad x \in X,$$

another application of Corollary 2.10 shows that fg belongs to $M(X, \mathbf{X})$.

It has been seen that the limit of a sequence of functions in $M(X, \mathbf{X})$ belongs to $M(X, \mathbf{X})$. We shall now prove that a nonnegative function f in $M(X, \mathbf{X})$ is the limit of a monotone increasing sequence (φ_n) in $M(X, \mathbf{X})$. Moreover, each φ_n can be chosen to be nonnegative and to assume only a finite number of real values.

2.11 LEMMA. *If f is a nonnegative function in $M(X, \mathbf{X})$, then there exists a sequence (φ_n) in $M(X, \mathbf{X})$ such that*

(a) $0 \leqslant \varphi_n(x) \leqslant \varphi_{n+1}(x)$ *for* $x \in X$, $n \in \mathbf{N}$.
(b) $f(x) = \lim \varphi_n(x)$ *for each* $x \in X$.
(c) *Each* φ_n *has only a finite number of real values.*

PROOF. Let n be a fixed natural number. If $k = 0, 1, \ldots, n2^n - 1$, let E_{kn} be the set

$$E_{kn} = \{x \in X : k2^{-n} \leqslant f(x) < (k + 1)2^{-n}\},$$

and if $k = n2^n$, let E_{kn} be the set $\{x \in X : f(x) \geqslant n\}$. We observe that the sets $\{E_{kn} : k = 0, 1, \ldots, n2^n\}$ are disjoint, belong to $\mathbf{X}$, and have union equal to X. If we define φ_n to be equal to $k2^{-n}$ on E_{kn}, then φ_n belongs to $M(X, \mathbf{X})$. It is readily established that the properties (a), (b), (c) hold. Q.E.D.

COMPLEX-VALUED FUNCTIONS

It is frequently important to consider complex-valued functions defined on X and to have a notion of measurability for such functions. We observe that if f is a complex-valued function defined on X, then there exist two uniquely determined real-valued functions f_1, f_2 such that

$$f = f_1 + if_2.$$

(Indeed, $f_1(x) = \operatorname{Re} f(x)$, $f_2(x) = \operatorname{Im} f(x)$, for $x \in X$.) We define the complex-valued function f to be **measurable** if and only if its **real** and **imaginary parts** f_1 and f_2, respectively, are measurable. It is easy to see that sums, products, and limits of complex-valued measurable functions are also measurable.

FUNCTIONS BETWEEN MEASURABLE SPACES

In the sequel we shall require the notion of measurability only for real- and complex-valued functions. In some work, however, one wishes to define measurability for a function f from one measurable space $(X, \boldsymbol{X})$ into another measurable space $(Y, \boldsymbol{Y})$. In this case one says that f is **measurable** in case the set

$$f^{-1}(E) = \{x \in X : f(x) \in E\}$$

belongs to $\boldsymbol{X}$ for every set E belonging to $\boldsymbol{Y}$. Although this definition of measurability appears to differ from Definition 2.3, it is not difficult to show (see Exercise 2.P) that Definition 2.3 is equivalent to this definition in the case that $Y = \boldsymbol{R}$ and $\boldsymbol{Y} = \boldsymbol{B}$.

This definition of measurability shows very clearly the close analogy between the measurable functions on a measurable space and continuous functions on a topological space.

EXERCISES

2.A. Show that $[a, b] = \bigcap_{n=1}^{\infty} (a - 1/n, b + 1/n)$. Hence any σ-algebra of subsets of $\boldsymbol{R}$ which contains all open intervals also contains all closed intervals. Similarly, $(a, b) = \bigcup_{n=1}^{\infty} [a + 1/n, b - 1/n]$, so that any σ-algebra containing all closed intervals also contains open intervals.

2.B. Show that the Borel algebra $\boldsymbol{B}$ is also generated by the collection of all half-open intervals $(a, b] = \{x \in \boldsymbol{R} : a < x \leqslant b\}$. Also show that $\boldsymbol{B}$ is generated by the collection of all half-rays $\{x \in \boldsymbol{R} : x > a\}$, $a \in \boldsymbol{R}$.

2.C. Let (A_n) be a sequence of subsets of a set X. Let $E_0 = \emptyset$ and for $n \in \boldsymbol{N}$, let

$$E_n = \bigcup_{k=1}^{n} A_k, \quad F_n = A_n \setminus E_{n-1}.$$

Show that (E_n) is a monotone increasing sequence of sets and that (F_n) is a disjoint sequence of sets (that is, $F_n \cap F_m = \emptyset$ if $n \neq m$) such that

$$\bigcup_{n=1}^{\infty} E_n = \bigcup_{n=1}^{\infty} F_n = \bigcup_{n=1}^{\infty} A_n.$$

2.D. Let (A_n) be a sequence of subsets of a set X. If A consists of all $x \in X$ which belong to infinitely many of the sets A_n, show that

$$A = \bigcap_{m=1}^{\infty} \left[\bigcup_{n=m}^{\infty} A_n \right].$$

The set A is often called the **limit superior** of the sets (A_n) and denoted by $\limsup A_n$.

2.E. Let (A_n) be a sequence of subsets of a set X. If B consists of all $x \in X$ which belong to all but a finite number of the sets A_n, show that

$$B = \bigcup_{m=1}^{\infty} \left[\bigcap_{n=m}^{\infty} A_n \right].$$

The set B is often called the **limit inferior** of the sets (A_n) and denoted by $\liminf A_n$.

2.F. If (E_n) is a sequence of subsets of a set X which is monotone increasing (that is, $E_1 \subseteq E_2 \subseteq E_3 \subseteq \cdots$), show that

$$\limsup E_n = \bigcup_{n=1}^{\infty} E_n = \liminf E_n.$$

2.G. If (F_n) is a sequence of subsets of a set X which is monotone decreasing (that is, $F_1 \supseteq F_2 \supseteq F_3 \supseteq \cdots$), show that

$$\limsup F_n = \bigcap_{n=1}^{\infty} F_n = \liminf F_n.$$

2.H. If (A_n) is a sequence of subsets of X, show that

$$\emptyset \subseteq \liminf A_n \subseteq \limsup A_n \subseteq X.$$

Give an example of a sequence (A_n) such that

$$\liminf A_n = \emptyset, \qquad \limsup A_n = X.$$

Give an example of a sequence (A_n) which is neither monotone increasing or decreasing, but is such that

$$\liminf A_n = \limsup A_n.$$

When this equality holds, the common value is called the **limit** of (A_n) and is denoted by $\lim A_n$.

2.I. Give an example of a function f on X to $\boldsymbol{R}$ which is not $\boldsymbol{X}$-measurable, but is such that the functions $|f|$ and f^2 are $\boldsymbol{X}$-measurable.

2.J. If a, b, c are real numbers, let mid (a, b, c) denote the "value in the middle." Show that

$$\text{mid}\,(a, b, c) = \inf \{\sup \{a, b\}, \sup \{a, c\}, \sup \{b, c\}\}.$$

If f_1, f_2, f_3 are $\boldsymbol{X}$-measurable functions on X to $\boldsymbol{R}$ and if g is defined for $x \in X$ by

$$g(x) = \text{mid}\,(f_1(x), \quad f_2(x), \quad f_3(x)),$$

then g is $\boldsymbol{X}$-measurable.

2.K. Show directly (without using the preceding exercise) that if f is measurable and $A > 0$, then the **truncation** f_A defined by

$$\begin{aligned} f_A(x) &= f(x), &&\text{if} \quad |f(x)| \leqslant A, \\ &= A, &&\text{if} \quad f(x) > A, \\ &= -A, &&\text{if} \quad f(x) < -A, \end{aligned}$$

is measurable.

2.L. Let f be a nonnegative $\boldsymbol{X}$-measurable function on X which is bounded (that is, there exists a constant K such that $0 \leqslant f(x) \leqslant K$ for all x in X). Show that the sequence (φ_n) constructed in Lemma 2.11 converges *uniformly* on X to f.

2.M. Let f be a function defined on a set X with values in a set Y. If E is any subset of Y, let

$$f^{-1}(E) = \{x \in X : f(x) \in E\}.$$

Show that $f^{-1}(\emptyset) = \emptyset$, $f^{-1}(Y) = X$. If E and F are subsets of Y, then

$$f^{-1}(E \setminus F) = f^{-1}(E) \setminus f^{-1}(F).$$

If $\{E_\alpha\}$ is any nonempty collection of subsets of Y, then

$$f^{-1}\Big(\bigcup_\alpha E_\alpha\Big) = \bigcup_\alpha f^{-1}(E_\alpha), \quad f^{-1}\Big(\bigcap_\alpha E_\alpha\Big) = \bigcap_\alpha f^{-1}(E_\alpha).$$

In particular it follows that if $\boldsymbol{Y}$ is a σ-algebra of subsets of Y, then $\{f^{-1}(E) : E \in \boldsymbol{Y}\}$ is a σ-algebra of subsets of X.

2.N. Let f be a function defined on a set X with values in a set Y. Let $\boldsymbol{X}$ be a σ-algebra of subsets of X and let $\boldsymbol{Y} = \{E \subseteq Y : f^{-1}(E) \in \boldsymbol{X}\}$. Show that $\boldsymbol{Y}$ is a σ-algebra.

2.O. Let $(X, \boldsymbol{X})$ be a measurable space and f be defined on X to Y. Let $\boldsymbol{A}$ be a collection of subsets of Y such that $f^{-1}(E) \in \boldsymbol{X}$ for every $E \in \boldsymbol{A}$. Show that $f^{-1}(F) \in \boldsymbol{X}$ for any set F which belongs to the σ-algebra generated by $\boldsymbol{A}$. (*Hint:* Use the preceding exercise.)

2.P. Let $(X, \boldsymbol{X})$ be a measurable space and f be a real-valued function defined on X. Show that f is $\boldsymbol{X}$-measurable if and only if $f^{-1}(E) \in \boldsymbol{X}$ for every Borel set E.

2.Q. Let $(X, \boldsymbol{X})$ be a measurable space, f be an $\boldsymbol{X}$-measurable function on X to $\boldsymbol{R}$ and let φ be a continuous function on $\boldsymbol{R}$ to $\boldsymbol{R}$. Show that the composition $\varphi \circ f$, defined by $(\varphi \circ f)(x) = \varphi[f(x)]$, is $\boldsymbol{X}$-measurable. (*Hint:* If φ is continuous, then $\varphi^{-1}(E) \in \boldsymbol{B}$ for each $E \in \boldsymbol{B}$.)

2.R. Let f be as in the preceding exercise and let ψ be a Borel measurable function. Show that $\psi \circ f$ is $\boldsymbol{X}$-measurable.

2.S. Let f be a complex-valued function defined on a measurable space $(X, \boldsymbol{X})$. Show that f is $\boldsymbol{X}$-measurable if and only if

$$\{x \in X : a < \operatorname{Re} f(x) < b, \quad c < \operatorname{Im} f(x) < d\}$$

belongs to $\boldsymbol{X}$ for all real numbers a, b, c, d. More generally, f is $\boldsymbol{X}$-measurable if and only if $f^{-1}(G) \in \boldsymbol{X}$ for every open set G in the complex plane $\boldsymbol{C}$.

2.T. Show that sums, products, and limits of complex-valued measurable functions are measurable.

2.U. Show that a function f on X to $\boldsymbol{R}$ (or to $\bar{\boldsymbol{R}}$) is $\boldsymbol{X}$-measurable if and only if the set A_α in Lemma 2.4(a) belongs to $\boldsymbol{X}$ for each rational number α; or, if and only if the set B_α in Lemma 2.4(b) belongs to $\boldsymbol{X}$ for each rational number α; etc.

2.V. A nonempty collection $\boldsymbol{M}$ of subsets of a set X is called a **monotone class** if, for each monotone increasing sequence (E_n) in $\boldsymbol{M}$ and each monotone decreasing sequence (F_n) in $\boldsymbol{M}$, the sets

$$\bigcup_{n=1}^{\infty} E_n, \qquad \bigcap_{n=1}^{\infty} F_n$$

belong to $\boldsymbol{M}$. Show that a σ-algebra is a monotone class. Also, if $\boldsymbol{A}$ is a nonempty collection of subsets of X, then there is a smallest monotone class containing $\boldsymbol{A}$. (This smallest monotone class is called the **monotone class generated by** $\boldsymbol{A}$.)

2.W. If $\boldsymbol{A}$ is a nonempty collection of subsets of X, then the σ-algebra $\boldsymbol{S}$ generated by $\boldsymbol{A}$ contains the monotone class $\boldsymbol{M}$ generated by $\boldsymbol{A}$. Show that the inclusion $\boldsymbol{A} \subseteq \boldsymbol{M} \subseteq \boldsymbol{S}$ may be proper.

CHAPTER 3

Measures

We have introduced the notion of a measurable space $(X, \boldsymbol{X})$ consisting of a set X and a σ-algebra $\boldsymbol{X}$ of subsets of X. We now consider certain functions which are defined on $\boldsymbol{X}$ and have real, or extended real values. These functions, which will be called "measures," are suggested by our idea of length, area, mass, and so forth. Thus it is natural that they should attach the value 0 to the empty set $\emptyset$ and that they should be additive over disjoint sets in $\boldsymbol{X}$. (Actually we shall require that they be countably additive in the sense to be described below.) It is also desirable to permit the measures to take on the extended real number $+\infty$.

3.1 DEFINITION. A **measure** is an extended real-valued function μ defined on a σ-algebra $\boldsymbol{X}$ of subsets of X such that (i) $\mu(\emptyset) = 0$, (ii) $\mu(E) \geqslant 0$ for all $E \in \boldsymbol{X}$, and (iii) μ is **countably additive** in the sense that if (E_n) is any disjoint sequence* of sets in $\boldsymbol{X}$, then

$$\mu\left(\bigcup_{n=1}^{\infty} E_n\right) = \sum_{n=1}^{\infty} \mu(E_n). \tag{3.1}$$

Since we permit μ to take on $+\infty$, we remark that the appearance of the value $+\infty$ on the right side of the equation (3.1) means either that $\mu(E_n) = +\infty$ for some n or that the series of nonnegative terms on the right side of (3.1) is divergent. If a measure does not take on $+\infty$,

* This means that $E_n \cap E_m = \emptyset$ if $n \neq m$.

we say that it is **finite**. More generally, if there exists a sequence (E_n) of sets in $\boldsymbol{X}$ with $X = \bigcup E_n$ and such that $\mu(E_n) < +\infty$ for all n, then we say that μ is **σ-finite.**

3.2 EXAMPLES. (a) Let X be any nonempty set and let $\boldsymbol{X}$ be the σ-algebra of all subsets of X. Let μ_1 be defined on $\boldsymbol{X}$ by

$$\mu_1(E) = 0, \qquad \text{for all } E \in \boldsymbol{X};$$

and let μ_2 be defined by

$$\mu_2(\emptyset) = 0, \quad \mu_2(E) = +\infty \qquad \text{if} \quad E \neq \emptyset.$$

Both μ_1 and μ_2 are measures, although neither one is very interesting. Note that μ_2 is neither finite nor σ-finite.

(b) Let $(X, \boldsymbol{X})$ be as in (a) and let p be a fixed element of X. Let μ be defined for $E \in \boldsymbol{X}$ by

$$\begin{aligned} \mu(E) &= 0, \qquad \text{if} \quad p \notin E, \\ &= 1, \qquad \text{if} \quad p \in E. \end{aligned}$$

It is readily seen that μ is a finite measure; it is called the **unit measure concentrated at** p.

(c) Let $X = \boldsymbol{N} = \{1, 2, 3, \ldots\}$ and let $\boldsymbol{X}$ be the σ-algebra of all subsets of $\boldsymbol{N}$. If $E \in \boldsymbol{X}$, define $\mu(E)$ to be equal to the number of elements in E if E is a finite set and equal to $+\infty$ if E is an infinite set. Then μ is a measure and is called the **counting measure on** $\boldsymbol{N}$. Note that μ is not finite, but it is σ-finite.

(d) If $X = \boldsymbol{R}$ and $\boldsymbol{X} = \boldsymbol{B}$, the Borel algebra, then it will be shown in Chapter 9 that there exists a unique measure λ defined on $\boldsymbol{B}$ which coincides with length on open intervals. [By this we mean that if E is the nonempty interval (a, b), then $\lambda(E) = b - a$.] This unique measure is usually called **Lebesgue** (or **Borel**) **measure**. It is not a finite measure, but it is σ-finite.

(e) If $X = \boldsymbol{R}$, $\boldsymbol{X} = \boldsymbol{B}$, and f is a continuous monotone increasing function, then it will be shown in Chapter 9 that there exists a unique measure λ_f defined on $\boldsymbol{B}$ such that if $E = (a, b)$, then $\lambda_f(E) = f(b) - f(a)$. This measure λ_f is called the **Borel-Stieltjes measure generated by** f.

We shall now derive a few simple results that will be needed later.

3.3 LEMMA. *Let μ be a measure defined on a σ-algebra $\boldsymbol{X}$. If E and F belong to $\boldsymbol{X}$ and $E \subseteq F$, then $\mu(E) \leqslant \mu(F)$. If $\mu(E) < +\infty$, then $\mu(F \setminus E) = \mu(F) - \mu(E)$.*

PROOF. Since $F = E \cup (F \setminus E)$ and $E \cap (F \setminus E) = \emptyset$, it follows that

$$\mu(F) = \mu(E) + \mu(F \setminus E).$$

Since $\mu(F \setminus E) \geqslant 0$, it follows that $\mu(F) \geqslant \mu(E)$. If $\mu(E) < +\infty$, then we can subtract it from both sides of this equation. Q.E.D.

3.4 LEMMA. *Let μ be a measure defined on a σ-algebra $\boldsymbol{X}$.*

(a) *If (E_n) is an increasing sequence in $\boldsymbol{X}$, then*

$$\mu\left(\bigcup_{n=1}^{\infty} E_n\right) = \lim \mu(E_n). \tag{3.2}$$

(b) *If (F_n) is a decreasing sequence in $\boldsymbol{X}$ and if $\mu(F_1) < +\infty$, then*

$$\mu\left(\bigcap_{n=1}^{\infty} F_n\right) = \lim \mu(F_n). \tag{3.3}$$

PROOF. (a) If $\mu(E_n) = +\infty$ for some n, then both sides of equation (3.2) are $+\infty$. Hence we can suppose that $\mu(E_n) < +\infty$ for all n.
Let $A_1 = E_1$ and $A_n = E_n \setminus E_{n-1}$ for $n > 1$. Then (A_n) is a disjoint sequence of sets in $\boldsymbol{X}$ such that

$$E_n = \bigcup_{j=1}^{n} A_j, \qquad \bigcup_{n=1}^{\infty} E_n = \bigcup_{n=1}^{\infty} A_n.$$

Since μ is countably additive,

$$\mu\left(\bigcup_{n=1}^{\infty} E_n\right) = \sum_{n=1}^{\infty} \mu(A_n) = \lim \sum_{n=1}^{m} \mu(A_n).$$

By Lemma 3.3 $\mu(A_n) = \mu(E_n) - \mu(E_{n-1})$ for $n > 1$, so the finite series on the right side telescopes and

$$\sum_{n=1}^{m} \mu(A_n) = \mu(E_m).$$

Hence equation (3.2) is proved.

(b) Let $E_n = F_1 \setminus F_n$, so that (E_n) is an increasing sequence of sets in $\boldsymbol{X}$. If we apply part (a) and Lemma 3.3, we infer that

$$\mu\left(\bigcup_{n=1}^{\infty} E_n\right) = \lim \mu(E_n) = \lim [\mu(F_1) - \mu(F_n)]$$
$$= \mu(F_1) - \lim \mu(F_n).$$

Since $\bigcup_{n=1}^{\infty} E_n = F_1 \setminus \bigcap_{n=1}^{\infty} F_n$, it follows that

$$\mu\left(\bigcup_{n=1}^{\infty} E_n\right) = \mu(F_1) - \mu\left(\bigcap_{n=1}^{\infty} F_n\right).$$

Combining these two equations, we obtain (3.3). Q.E.D.

3.5 Definition. A **measure space** is a triple $(X, \boldsymbol{X}, \mu)$ consisting of a set X, a σ-algebra $\boldsymbol{X}$ of subsets of X, and a measure μ defined on $\boldsymbol{X}$.

There is a terminological matter that needs to be mentioned and which shall be employed in the following. We shall say that a certain proposition holds **μ-almost everywhere** if there exists a subset $N \in \boldsymbol{X}$ with $\mu(N) = 0$ such that the proposition holds on the complement of N. Thus we say that two functions f, g are **equal μ-almost everywhere** or that they are **equal for μ-almost all** x in case $f(x) = g(x)$ when $x \notin N$, for some $N \in \boldsymbol{X}$ with $\mu(N) = 0$. In this case we will often write

$$f = g, \quad \mu\text{-a.e.}$$

In like manner, we say that a sequence (f_n) of functions on X **converges μ-almost everywhere** (or **converges for μ-almost all** x) if there exists a set $N \in \boldsymbol{X}$ with $\mu(N) = 0$ such that $f(x) = \lim f_n(x)$ for $x \notin N$. In this case we often write

$$f = \lim f_n, \quad \mu\text{-a.e.}$$

Of course, if the measure μ is understood, we shall say "almost everywhere" instead of "μ-almost everywhere."

There are some instances (suggested by the notion of electrical charge, for example) in which it is desirable to discuss functions which behave like measures except that they take both positive and negative values. In this case, it is not so convenient to permit the extended real numbers $+\infty, -\infty$ to be values since we wish to avoid expressions of the form $(+\infty) + (-\infty)$. Although it is possible to handle "signed measures" which take on only *one* of the values $+\infty$, $-\infty$, we shall restrict our attention to the case where neither of these symbols is permitted. To indicate this restriction, we shall introduce the term "charge," which is not entirely standard.

3.6 DEFINITION. If $\boldsymbol{X}$ is a σ-algebra of subsets of a set X, then a real-valued function λ defined on $\boldsymbol{X}$ is said to be a **charge** in case $\lambda(\emptyset) = 0$ and λ is countably additive in the sense that if (E_n) is a disjoint sequence of sets in $\boldsymbol{X}$, then

$$\lambda\left(\bigcup_{n=1}^{\infty} E_n\right) = \sum_{n=1}^{\infty} \lambda(E_n).$$

[Since the left-hand side is independent of the order and this equality is required for all such sequences, the series on the right-hand side must be unconditionally convergent for all disjoint sequences of measurable sets.]

It is clear that the sum and difference of two charges is a charge. More generally, any finite linear combination of charges is a charge. It will be seen in Chapter 5 that functions which are integrable over a measure space $(X, \boldsymbol{X}, \mu)$ give rise to charges. Later, in Chapter 8, we will characterize those charges which are generated by integrable functions.

EXERCISES

3.A. If μ is a measure on $\boldsymbol{X}$ and A is a fixed set in $\boldsymbol{X}$, then the function λ, defined for $E \in \boldsymbol{X}$ by $\lambda(E) = \mu(A \cap E)$, is a measure on $\boldsymbol{X}$.

3.B. If $\mu_1, \ldots, \mu_n$ are measures on $\boldsymbol{X}$ and $a_1, \ldots, a_n$ are nonnegative real numbers, then the function λ, defined for $E \in \boldsymbol{X}$ by

$$\lambda(E) = \sum_{j=1}^{n} a_j\, \mu_j(E),$$

is a measure on $\boldsymbol{X}$.

3.C. If (μ_n) is a sequence of measures on $\boldsymbol{X}$ with $\mu_n(X) = 1$ and if λ is defined by

$$\lambda(E) = \sum_{n=1}^{\infty} 2^{-n}\, \mu_n(E), \quad E \in \boldsymbol{X},$$

then λ is a measure on $\mathbf{X}$ and $\lambda(X) = 1$.

3.D. Let $X = \mathbf{N}$ and let $\mathbf{X}$ be the σ-algebra of all subsets of $\mathbf{N}$. If (a_n) is a sequence of nonnegative real numbers and if we define μ by

$$\mu(\emptyset) = 0; \quad \mu(E) = \sum_{n \in E} a_n, \quad E \neq \emptyset;$$

then μ is a measure on $\mathbf{X}$. Conversely, every measure on $\mathbf{X}$ is obtained in this way for some sequence (a_n) in $\overline{\mathbf{R}}$.

3.E. Let X be an uncountable set and let $\mathbf{X}$ be the family of all subsets of X. Define μ on E in $\mathbf{X}$ by requiring that $\mu(E) = 0$, if E is countable, and $\mu(E) = +\infty$, if E is uncountable. Show that μ is a measure on $\mathbf{X}$.

3.F. Let $X = \mathbf{N}$ and let $\mathbf{X}$ be the family of all subsets of $\mathbf{N}$. If E is finite, let $\mu(E) = 0$; if E is infinite, let $\mu(E) = +\infty$. Is μ a measure on $\mathbf{X}$?

3.G. If X and $\mathbf{X}$ are as in Exercise 3.F, let $\lambda(E) = +\infty$ for all $E \in \mathbf{X}$. Is λ a measure?

3.H. Show that Lemma 3.4(b) may fail if the finiteness condition $\mu(F_1) < +\infty$ is dropped.

3.I. Let $(X, \mathbf{X}, \mu)$ be a measure space and let (E_n) be a sequence in $\mathbf{X}$. Show that

$$\mu(\liminf E_n) \leqslant \liminf \mu(E_n).$$

[See Exercise 2.E.]

3.J. Using the notation of Exercise 2.D, show that

$$\limsup \mu(E_n) \leqslant \mu(\limsup E_n)$$

when $\mu(\bigcup E_n) < +\infty$. Show that this inequality may fail if $\mu(\bigcup E_n) = +\infty$.

3.K. Let $(X, \mathbf{X}, \mu)$ be a measure space and let $\mathbf{Z} = \{E \in \mathbf{X} : \mu(E) = 0\}$. Is $\mathbf{Z}$ a σ-algebra? Show that if $E \in \mathbf{Z}$ and $F \in \mathbf{X}$, then $E \cap F \in \mathbf{Z}$. Also, if E_n belongs to $\mathbf{Z}$ for $n \in \mathbf{N}$, then $\bigcup E_n \in \mathbf{Z}$.

3.L. Let $X, \mathbf{X}, \mu, \mathbf{Z}$ be as in Exercise 3.K and let $\mathbf{X}'$ be the family of all subsets of X of the form

$$(E \cup Z_1) \setminus Z_2, \quad E \in \mathbf{X},$$

where Z_1 and Z_2 are arbitrary subsets of sets belonging to $\boldsymbol{Z}$. Show that a set is in $\boldsymbol{X}'$ if and only if it has the form $E \cup Z$ where $E \in \boldsymbol{X}$ and Z is a subset of a set in $\boldsymbol{Z}$. Show that the collection $\boldsymbol{X}'$ forms a σ-algebra of sets in X. The σ-algebra $\boldsymbol{X}'$ is called the **completion** of $\boldsymbol{X}$ (with respect to μ).

3.M. With the notation of Exercise 3.L, let μ' be defined on $\boldsymbol{X}'$ by

$$\mu'(E \cup Z) = \mu(E),$$

when $E \in \boldsymbol{X}$ and Z is a subset of a set in $\boldsymbol{Z}$. Show that μ' is well-defined and is a measure on $\boldsymbol{X}'$ which agrees with μ on $\boldsymbol{X}$. The measure μ' is called the **completion** of μ.

3.N. Let $(X, \boldsymbol{X}, \mu)$ be a measure space and let $(X, \boldsymbol{X}', \mu')$ be its completion in the sense of Exercise 3.M. Suppose that f is an $\boldsymbol{X}'$-measurable function on X to $\bar{\boldsymbol{R}}$. Show that there exists an $\boldsymbol{X}$-measurable function g on X to $\bar{\boldsymbol{R}}$ which is μ-almost everywhere equal to f. (*Hint:* For each rational number r, let $A_r = \{x : f(x) > r\}$ and write $A_r = E_r \cup Z_r$, where $E_r \in \boldsymbol{X}$ and Z_r is a subset of a set in $\boldsymbol{Z}$. Let Z be a set in $\boldsymbol{Z}$ containing $\bigcup Z_r$ and define $g(x) = f(x)$ for $x \notin Z$, and $g(x) = 0$ for $x \in Z$. To show that g is $\boldsymbol{X}$-measurable, use Exercise 2.U.)

3.O. Show that Lemma 3.4 holds if μ is a charge on $\boldsymbol{X}$.

3.P. If μ is a charge on $\boldsymbol{X}$, let π be defined for $E \in \boldsymbol{X}$ by

$$\pi(E) = \sup \{\mu(A) : A \subseteq E, A \in \boldsymbol{X}\}.$$

Show that π is a measure on $\boldsymbol{X}$. (*Hint:* If $\pi(E_n) < \infty$ and $\varepsilon > 0$, let $F_n \in \boldsymbol{X}$ be such that $F_n \subseteq E_n$ and $\pi(E_n) \leqslant \mu(F_n) + 2^{-n}\varepsilon$.)

3.Q. If μ is a charge on $\boldsymbol{X}$, let ν be defined for $E \in \boldsymbol{X}$ by

$$\nu(E) = \sup \sum_{j=1}^{n} |\mu(A_j)|,$$

where the supremum is taken over all finite disjoint collections $\{A_j\}$ in $\boldsymbol{X}$ with $E = \bigcup_{j=1}^{n} A_j$. Show that ν is a measure on $\boldsymbol{X}$. (It is called the **variation of** μ.)

3.R. Let λ denote Lebesgue measure defined on the Borel algebra $\boldsymbol{B}$ of $\boldsymbol{R}$ [see Example 3.2(d)]. (a) If E consists of a single point, then $E \in \boldsymbol{B}$ and $\lambda(E) = 0$. (b) If E is countable, then $E \in \boldsymbol{B}$ and $\lambda(E) = 0$.

(c) The open interval (a, b), the half-open intervals $(a, b]$, $[a, b)$, and the closed interval $[a, b]$ all have Lebesgue measure $b - a$.

3.S. If λ denotes Lebesgue measure and E is an open subset of $\boldsymbol{R}$, then $\lambda(E) > 0$. Use the Heine–Borel Theorem (see Reference [1], p. 85) to show that if K is a compact subset of $\boldsymbol{R}$, then $\lambda(K) < +\infty$.

3.T. Show that the Lebesgue measure of the Cantor set (see Reference [1], p. 52) is zero.

3.U. By varying the construction of the Cantor set, obtain a set of positive Lebesgue measure which contains no nonvoid open interval.

3.V. Suppose that E is a subset of a set $N \in \boldsymbol{X}$ with $\mu(N) = 0$ but that $E \notin \boldsymbol{X}$. The sequence (f_n), $f_n = 0$, converges μ-almost everywhere to χ_E. Hence the almost everywhere limit of a sequence of measurable functions may not be measurable.

CHAPTER 4

The Integral

In this chapter we shall introduce the integral first for nonnegative simple measurable functions and then for arbitrary nonnegative extended real-valued measurable functions. The principal result is the celebrated Monotone Convergence Theorem, which is a basic tool for everything that follows.

Throughout this chapter we shall consider a fixed measure space $(X, \boldsymbol{X}, \mu)$. We shall denote the collection of all $\boldsymbol{X}$-measurable functions on X to $\bar{\boldsymbol{R}}$ by $M = M(X, \boldsymbol{X})$ and the collection of all nonnegative $\boldsymbol{X}$-measurable functions on X to $\bar{\boldsymbol{R}}$ by $M^+ = M^+(X, \boldsymbol{X})$. We shall define the integral of any function in M^+ with respect to the measure μ. In order to do so we shall find it convenient to introduce the notion of a simple function. It is convenient to require that simple functions have values in $\boldsymbol{R}$ rather than in $\bar{\boldsymbol{R}}$.

4.1 DEFINITION. A real-valued function is **simple** if it has only a finite number of values.

A simple measurable function φ can be represented in the form

$$\varphi = \sum_{j=1}^{n} a_j \chi_{E_j}, \tag{4.1}$$

where $a_j \in \boldsymbol{R}$ and χ_{E_j} is the characteristic function of a set E_j in $\boldsymbol{X}$. Among these representations for φ there is a unique **standard representation** characterized by the fact that the a_j are distinct and the E_j

are disjoint. Indeed, if $a_1, a_2, \ldots, a_n$ are the distinct values of φ and if $E_j = \{x \in X : \varphi(x) = a_j\}$, then the E_j are disjoint and $X = \bigcup_{j=1}^{n} E_j$. (Of course, if we do not require the a_j to be distinct, or the sets E_j to be disjoint, then a simple function has many representations as a linear combination of characteristic functions.)

4.2 DEFINITION. If φ is a simple function in $M^+(X, \mathbf{X})$ with the standard representation (4.1), we define the **integral** of φ with respect to μ to be the extended real number

$$\int \varphi \, d\mu = \sum_{j=1}^{n} a_j \, \mu(E_j). \tag{4.2}$$

In the expression (4.2) we employ the convention that $0(+\infty) = 0$ so the integral of the function identically 0 is equal to 0 whether the space has finite or infinite measure. It should be noted that the value of the integral of a simple function in M^+ is well-defined (although it may be $+\infty$) since all the a_j are nonnegative, and so we do not encounter meaningless expressions such as $(+\infty) - (+\infty)$.

We shall need the following elementary properties of the integral.

4.3 LEMMA. (a) *If φ and ψ are simple functions in $M^+(X, \mathbf{X})$ and $c \geqslant 0$, then*

$$\int c\varphi \, d\mu = c \int \varphi \, d\mu,$$

$$\int (\varphi + \psi) \, d\mu = \int \varphi \, d\mu + \int \psi \, d\mu.$$

(b) *If λ is defined for E in $\mathbf{X}$ by*

$$\lambda(E) = \int \varphi \, \chi_E \, d\mu,$$

then λ is a measure on $\mathbf{X}$.

PROOF. If $c = 0$, then $c\varphi$ vanishes identically and the equality holds. If $c > 0$, then $c\varphi$ is in M^+ with standard representation

$$c\varphi = \sum_{j=1}^{n} c a_j \, \chi_{E_j},$$

when φ has standard representation (4.1). Therefore

$$\int c\varphi \, d\mu = \sum_{j=1}^{n} ca_j \, \mu(E_j) = c \sum_{j=1}^{n} a_j \, \mu(E_j) = c \int \varphi \, d\mu.$$

Let φ and ψ have standard representations

$$\varphi = \sum_{j=1}^{n} a_j \, \chi_{E_j}, \qquad \psi = \sum_{k=1}^{m} b_k \, \chi_{F_k},$$

then $\varphi + \psi$ has a representation

$$\varphi + \psi = \sum_{j=1}^{n} \sum_{k=1}^{m} (a_j + b_k) \chi_{E_j \cap F_k}.$$

However, this representation of $\varphi + \psi$ as a linear combination of characteristic functions of the disjoint sets $E_j \cap F_k$ is not necessarily the standard representation for $\varphi + \psi$, since the values $a_j + b_k$ may not be distinct. Let c_h, $h = 1, \ldots, p$, be the distinct numbers in the set $\{a_j + b_k : j = 1, \ldots, n; k = 1, \ldots, m\}$ and let G_h be the union of all those sets $E_j \cap F_k$ such that $a_j + b_k = c_h$. Thus

$$\mu(G_h) = \sum_{(h)} \mu(E_j \cap F_k),$$

where the notation designates summation over all j, k such that $a_j + b_k = c_h$. Since the standard representation of $\varphi + \psi$ is given by

$$\varphi + \psi = \sum_{h=1}^{p} c_h \, \chi_{G_h},$$

we find that

$$\begin{aligned}
\int (\varphi + \psi) \, d\mu &= \sum_{h=1}^{p} c_h \, \mu(G_h) = \sum_{h=1}^{p} \sum_{(h)} c_h \, \mu(E_j \cap F_k) \\
&= \sum_{h=1}^{p} \sum_{(h)} (a_j + b_k) \, \mu(E_j \cap F_k) \\
&= \sum_{j=1}^{n} \sum_{k=1}^{m} (a_j + b_k) \, \mu(E_j \cap F_k) \\
&= \sum_{j=1}^{n} \sum_{k=1}^{m} a_j \, \mu(E_j \cap F_k) + \sum_{j=1}^{n} \sum_{k=1}^{m} b_k \, \mu(E_j \cap F_k).
\end{aligned}$$

Since X is the union of both of the disjoint families $\{E_j\}$ and $\{F_k\}$, then

$$\mu(E_j) = \sum_{k=1}^{m} \mu(E_j \cap F_k), \quad \mu(F_k) = \sum_{j=1}^{n} \mu(E_j \cap F_k).$$

We employ this observation (and change the order of summation in the second term) to obtain the desired relation

$$\begin{aligned} \int (\varphi + \psi)\, d\mu &= \sum_{j=1}^{n} a_j\, \mu(E_j) + \sum_{k=1}^{m} b_k\, \mu(F_k) \\ &= \int \varphi\, d\mu + \int \psi\, d\mu. \end{aligned}$$

To establish part (b), we observe that

$$\varphi\, \chi_E = \sum_{j=1}^{n} a_j\, \chi_{E_j \cap E}.$$

Hence, it follows by induction from what we have proved that

$$\lambda(E) = \int \varphi\, \chi_E\, d\mu = \sum_{j=1}^{n} a_j \int \chi_{E_j \cap E}\, d\mu = \sum_{j=1}^{n} a_j\, \mu(E_j \cap E).$$

Since the mapping $E \to \mu(E_j \cap E)$ is a measure (see Exercise 3.A) we have expressed λ as a linear combination of measures on $\boldsymbol{X}$. It follows (see Exercise 3.B) that λ is also a measure on $\boldsymbol{X}$. Q.E.D.

We are now prepared to introduce the integral of an arbitrary function in M^+. Observe that we do not require the value of the integral to be finite.

4.4 Definition. If f belongs to $M^+(X, \boldsymbol{X})$, we define the **integral of f with respect to μ** to be the extended real number

$$\int f\, d\mu = \sup \int \varphi\, d\mu, \tag{4.3}$$

where the supremum is extended over all simple functions φ in $M^+(X, \boldsymbol{X})$ satisfying $0 \leqslant \varphi(x) \leqslant f(x)$ for all $x \in X$. If f belongs to $M^+(X, \boldsymbol{X})$ and E belongs to $\boldsymbol{X}$, then $f\chi_E$ belongs to $M^+(X, \boldsymbol{X})$ and we define the **integral of f over E with respect to μ** to be the extended real number

$$\int_E f\, d\mu = \int f\, \chi_E\, d\mu. \tag{4.4}$$

We shall first show that the integral is monotone both with respect to the integrand and the set over which the integral is extended.

4.5 LEMMA. (a) *If f and g belong to $M^+(X, \mathbf{X})$ and $f \leqslant g$, then*

$$\int f\,d\mu \leqslant \int g\,d\mu. \tag{4.5}$$

(b) *If f belongs to $M^+(X, \mathbf{X})$, if E, F belong to $\mathbf{X}$, and if $E \subseteq F$, then*

$$\int_E f\,d\mu \leqslant \int_F f\,d\mu.$$

PROOF. (a) If φ is a simple function in M^+ such that $0 \leqslant \varphi \leqslant f$, then $0 \leqslant \varphi \leqslant g$. Therefore (4.5) holds.

(b) Since $f\chi_E \leqslant f\chi_F$, part (b) follows from (a). Q.E.D.

We are now prepared to establish an important result due to B. Levi. This theorem provides the key to the fundamental convergence properties of the Lebesgue integral.

4.6 MONOTONE CONVERGENCE THEOREM. *If (f_n) is a monotone increasing sequence of functions in $M^+(X, \mathbf{X})$ which converges to f, then*

$$\int f\,d\mu = \lim \int f_n\,d\mu. \tag{4.6}$$

know proof

PROOF. According to Corollary 2.10, the function f is measurable. Since $f_n \leqslant f_{n+1} \leqslant f$, it follows from Lemma 4.5(a) that

$$\int f_n\,d\mu \leqslant \int f_{n+1}\,d\mu \leqslant \int f\,d\mu$$

for all $n \in \mathbf{N}$. Therefore we have

$$\lim \int f_n\,d\mu \leqslant \int f\,d\mu.$$

To establish the opposite inequality, let α be a real number satisfying $0 < \alpha < 1$ and let φ be a simple measurable function satisfying $0 \leqslant \varphi \leqslant f$. Let

$$A_n = \{x \in X : f_n(x) \geqslant \alpha\varphi(x)\}$$

so that $A_n \in \boldsymbol{X}$, $A_n \subseteq A_{n+1}$, and $X = \bigcup A_n$. According to Lemma 4.5,

$$\text{(4.7)} \qquad \int_{A_n} \alpha\varphi \, d\mu \leqslant \int_{A_n} f_n \, d\mu \leqslant \int f_n \, d\mu.$$

Since the sequence (A_n) is monotone increasing and has union X, it follows from Lemmas 4.3(b) and 3.4(a) that

$$\int \varphi \, d\mu = \lim \int_{A_n} \varphi \, d\mu.$$

Therefore, on taking the limit in (4.7) with respect to n, we obtain

$$\alpha \int \varphi \, d\mu \leqslant \lim \int f_n \, d\mu.$$

Since this holds for all α with $0 < \alpha < 1$, we infer that

$$\int \varphi \, d\mu \leqslant \lim \int f_n \, d\mu$$

and since φ is an arbitrary simple function in M^+ satisfying $0 \leqslant \varphi \leqslant f$, we conclude that

$$\int f \, d\mu = \sup_{\varphi} \int \varphi \, d\mu \leqslant \lim \int f_n \, d\mu.$$

If we combine this with the opposite inequality, we obtain (4.6). Q.E.D.

REMARK. It should be observed that it is not being assumed that either side of (4.6) is finite. Indeed, the sequence $(\int f_n \, d\mu)$ is a monotone increasing sequence of extended real numbers and so always has a limit in $\bar{\boldsymbol{R}}$, but perhaps not in $\boldsymbol{R}$.

We shall now derive some consequences of the Monotone Convergence Theorem.

4.7 COROLLARY. (a) *If f belongs to M^+ and $c \geqslant 0$, then cf belongs to M^+ and*

$$\int cf \, d\mu = c \int f \, d\mu.$$

(b) *If f, g belong to M^+, then $f + g$ belongs to M^+ and*

$$\int (f + g) \, d\mu = \int f \, d\mu + \int g \, d\mu.$$

PROOF. (a) If $c = 0$ the result is immediate. If $c > 0$, let (φ_n) be a monotone increasing sequence of simple functions in M^+ converging to f on X (see Lemma 2.11). Then $(c\varphi_n)$ is a monotone sequence converging to cf. If we apply Lemma 4.3(a) and the Monotone Convergence Theorem, we obtain

$$\begin{aligned}\int cf\, d\mu &= \lim \int c\, \varphi_n\, d\mu \\ &= c \lim \int \varphi_n\, d\mu = c \int f\, d\mu.\end{aligned}$$

(b) If (φ_n) and (ψ_n) are monotone increasing sequences of simple functions converging to f and g, respectively, then $(\varphi_n + \psi_n)$ is a monotone increasing sequence converging to $f + g$. It follows from Lemma 4.3(a) and the Monotone Convergence Theorem that

$$\begin{aligned}\int (f + g)\, d\mu &= \lim \int (\varphi_n + \psi_n)\, d\mu \\ &= \lim \int \varphi_n\, d\mu + \lim \int \psi_n\, d\mu \\ &= \int f\, d\mu + \int g\, d\mu.\end{aligned}$$

Q.E.D.

The next result, a consequence of the Monotone Convergence Theorem, is very important for it enables us to handle sequences of functions that are not monotone.

4.8 FATOU'S LEMMA. *If (f_n) belongs to $M^+(X, \mathbf{X})$, then*

$$\int (\liminf f_n)\, d\mu \leqslant \liminf \int f_n\, d\mu. \tag{4.8}$$

PROOF. Let $g_m = \inf\{f_m, f_{m+1}, \ldots\}$ so that $g_m \leqslant f_n$ whenever $m \leqslant n$. Therefore

$$\int g_m\, d\mu \leqslant \int f_n\, d\mu, \qquad m \leqslant n,$$

so that

$$\int g_m\, d\mu \leqslant \liminf \int f_n\, d\mu.$$

Since the sequence (g_m) is increasing and converges to $\liminf f_n$, the

Monotone Convergence Theorem implies that

$$\int (\liminf f_n)\, d\mu = \lim \int g_m \, d\mu$$

$$\leqslant \liminf \int f_n \, d\mu. \qquad \text{Q.E.D.}$$

It will be seen in an exercise that Fatou's Lemma may fail if it is not assumed that $f_n \geqslant 0$.

4.9 COROLLARY. *If f belongs to M^+ and if λ is defined on $\boldsymbol{X}$ by*

$$\lambda(E) = \int_E f \, d\mu, \tag{4.9}$$

then λ is a measure.

PROOF. Since $f \geqslant 0$ it follows that $\lambda(E) \geqslant 0$. If $E = \emptyset$, then $f\chi_E$ vanishes everywhere so that $\lambda(\emptyset) = 0$. To see that λ is countably additive, let (E_n) be a disjoint sequence of sets in $\boldsymbol{X}$ with union E and let f_n be defined to be

$$f_n = \sum_{k=1}^{n} f\chi_{E_k}.$$

It follows from Corollary 4.7(b) and induction that

$$\int f_n \, d\mu = \sum_{k=1}^{n} \int f\chi_{E_k} \, d\mu = \sum_{k=1}^{n} \lambda(E_k).$$

Since (f_n) is an increasing sequence in M^+ converging to $f\chi_E$, the Monotone Convergence Theorem implies that

$$\lambda(E) = \int f\chi_E \, d\mu = \lim \int f_n \, d\mu = \sum_{k=1}^{\infty} \lambda(E_k). \qquad \text{Q.E.D.}$$

4.10 COROLLARY. *Suppose that f belongs to M^+. Then $f(x) = 0$ μ-almost everywhere on X if and only if*

$$\int f \, d\mu = 0. \tag{4.10}$$

PROOF. If equation (4.10) holds, let

$$E_n = \left\{ x \in X : f(x) > \frac{1}{n} \right\},$$

so that $f \geqslant (1/n)\,\chi_{E_n}$, from which

$$0 = \int f\,d\mu \geqslant \frac{1}{n}\,\mu(E_n) \geqslant 0.$$

It follows that $\mu(E_n) = 0$; hence the set

$$\{x \in X : f(x) > 0\} = \bigcup_{n=1}^{\infty} E_n$$

also has measure 0.

Conversely, let $f(x) = 0$ μ-almost everywhere. If

$$E = \{x \in X : f(x) > 0\},$$

then $\mu(E) = 0$. Let $f_n = n\,\chi_E$. Since $f \leqslant \liminf f_n$, it follows from Fatou's Lemma that

$$0 \leqslant \int f\,d\mu \leqslant \liminf \int f_n\,d\mu = 0. \qquad \text{Q.E.D.}$$

4.11 COROLLARY. *Suppose that f belongs to M^+, and define λ on $\mathbf{X}$ by equation (4.9). Then the measure λ is absolutely continuous with respect to μ in the sense that if $E \in \mathbf{X}$ and $\mu(E) = 0$, then $\lambda(E) = 0$.*

PROOF. If $\mu(E) = 0$ for some $E \in \mathbf{X}$, then $f\chi_E$ vanishes μ-almost everywhere. By Corollary 4.10, we have

$$\lambda(E) = \int f\,\chi_E\,d\mu = 0. \qquad \text{Q.E.D.}$$

We shall now show that the Monotone Convergence Theorem holds if convergence on X is replaced by almost everywhere convergence.

4.12 COROLLARY. *If (f_n) is a monotone increasing sequence of functions in $M^+(X, \mathbf{X})$ which converges μ-almost everywhere on X to a function f in M^+, then*

$$\int f\,d\mu = \lim \int f_n\,d\mu.$$

PROOF. Let $N \in \mathbf{X}$ be such that $\mu(N) = 0$ and (f_n) converges to f at every point of $M = X \setminus N$. Then $(f_n\,\chi_M)$ converges to $f\,\chi_M$ on X, so the Monotone Convergence Theorem implies that

$$\int f\,\chi_M\,d\mu = \lim \int f_n\,\chi_M\,d\mu.$$

Since $\mu(N) = 0$, the functions $f\chi_N$ and $f_n \chi_N$ vanish μ-almost everywhere. It follows from Corollary 4.10 that

$$\int f\chi_N \, d\mu = 0, \qquad \int f_n \chi_N \, d\mu = 0.$$

Since $f = f\chi_M + f\chi_N$ and $f_n = f_n \chi_M + f_n \chi_N$, it follows that

$$\int f \, d\mu = \int f\chi_M \, d\mu = \lim \int f_n \chi_M \, d\mu = \lim \int f_n \, d\mu. \qquad \text{Q.E.D.}$$

4.13 COROLLARY. *Let* (g_n) *be a sequence in* M^+, *then*

$$\int \left(\sum_{n=1}^{\infty} g_n \right) d\mu = \sum_{n=1}^{\infty} \left(\int g_n \, d\mu \right).$$

PROOF. Let $f_n = g_1 + \cdots + g_n$, and apply the Monotone Convergence Theorem. Q.E.D.

EXERCISES

4.A. If the simple function φ in $M^+(X, \boldsymbol{X})$ has the (not necessarily standard) representation

$$\varphi = \sum_{k=1}^{m} b_k \chi_{E_k},$$

where $b_k \in \boldsymbol{R}$ and $F_k \in \boldsymbol{X}$, show that

$$\int \varphi \, d\mu = \sum_{k=1}^{m} b_k \, \mu(F_k).$$

4.B. The sum, scalar multiple, and product of simple functions are simple functions. [In other words, the simple functions in $M(X, \boldsymbol{X})$ form a vector subspace of $M(X, \boldsymbol{X})$.]

4.C. If φ_1 and φ_2 are simple functions in $M(X, \boldsymbol{X})$, then

$$\psi = \sup\{\varphi_1, \varphi_2\}, \qquad \omega = \inf\{\varphi_1, \varphi_2\}$$

are also simple functions in $M(X, \boldsymbol{X})$.

4.D. If $f \in M^+$ and $c > 0$, then the mapping $\varphi \rightarrow \psi = c\varphi$ is a one-one correspondence between simple function φ in M^+ with $\varphi \leq f$

and simple functions ψ in M^+ with $\psi \leqslant cf$. Use this observation to give a different proof of Corollary 4.7(a).

4.E. Let f, g belong to M^+, let φ be a simple function in M^+ with $\varphi \leqslant f$, and let ω be a simple function in M^+ with $\omega \leqslant f + g$. Put $\theta_1 = \inf\{\omega, \varphi\}$ and $\theta_2 = \sup\{\omega - \varphi, 0\}$. Show that $\omega = \theta_1 + \theta_2$ and that $\theta_1 \leqslant f$ and $\theta_2 \leqslant g$.

4.F. Employ Exercise 4.E to establish Corollary 4.7(b) without using the Monotone Convergence Theorem.

4.G. Let $X = \boldsymbol{N}$, let $\boldsymbol{X}$ be all subsets of $\boldsymbol{N}$, and let μ be the counting measure on $\boldsymbol{X}$. If f is a nonnegative function on $\boldsymbol{N}$, then $f \in M^+(X, \boldsymbol{X})$ and

$$\int f \, d\mu = \sum_{n=1}^{\infty} f(n).$$

4.H. Let $X = \boldsymbol{R}$, $\boldsymbol{X} = \boldsymbol{B}$, and let λ be the Lebesgue measure on $\boldsymbol{B}$. If $f_n = \chi_{[0,n]}$, then the sequence is monotone increasing to $f = \chi_{[0,+\infty)}$. Although the functions are uniformly bounded by 1 and the integrals of the f_n are all finite, we have

$$\int f \, d\lambda = +\infty.$$

Does the Monotone Convergence Theorem apply?

4.I. Let $X = \boldsymbol{R}$, $\boldsymbol{X} = \boldsymbol{B}$, and λ be Lebesgue measure on $\boldsymbol{X}$. If $f_n = (1/n)\, \chi_{[n,+\infty)}$, then the sequence (f_n) is monotone decreasing and converges uniformly to $f = 0$, but

$$0 = \int f \, d\lambda \neq \lim \int f_n \, d\lambda = +\infty.$$

(Hence there is no theorem corresponding to the Monotone Convergence Theorem for a decreasing sequence in M^+.)

4.J.(a) Let $f_n = (1/n)\, \chi_{[0,n]}$, $f = 0$. Show that the sequence (f_n) converges uniformly to f, but that

$$\int f \, d\lambda \neq \lim \int f_n \, d\lambda.$$

Why does this not contradict the Monotone Convergence Theorem?

Does Fatou's Lemma apply?

(b) Let $g_n = n\,\chi_{[1/n,\,2/n]}$, $g = 0$. Show that

$$\int g\,d\lambda \neq \lim \int g_n\,d\lambda.$$

Does the sequence (g_n) converge uniformly to g? Does the Monotone Convergence Theorem apply? Does Fatou's Lemma apply?

4.K. If $(X, \boldsymbol{X}, \mu)$ is a finite measure space, and if (f_n) is a real-valued sequence in $M^+(X, \boldsymbol{X})$ which converges uniformly to a function f, then f belongs to $M^+(X, \boldsymbol{X})$, and

$$\int f\,d\mu = \lim \int f_n\,d\mu.$$

4.L. Let X be a finite closed interval $[a, b]$ in $\boldsymbol{R}$, let $\boldsymbol{X}$ be the collection of Borel sets in X, and let λ be Lebesgue measure on $\boldsymbol{X}$. If f is a nonnegative continuous function on X, show that

$$\int f\,d\lambda = \int_a^b f(x)\,dx,$$

where the right side denotes the Riemann integral of f. (*Hint:* First establish this equality for a nonnegative **step function**, that is, a linear combination of characteristic functions of intervals.)

4.M. Let $X = [0, +\infty)$, let $\boldsymbol{X}$ be the Borel subsets of X, and let λ be Lebesgue measure on $\boldsymbol{X}$. If f is a nonnegative continuous function on X, show that

$$\int f\,d\lambda = \lim_{b \to +\infty} \int_0^b f(x)\,dx.$$

Hence, if f is a nonnegative continuous function, the Lebesgue and the improper Riemann integrals coincide.

[The next three exercises deal with the integration of functions which do not belong to M^+. They can be omitted until the next chapter has been read. However, we include them here because they illustrate the restrictions required by Fatou's Lemma.]

4.N. If $f_n = (-1/n)\,\chi_{[0,\,n]}$, then the sequence (f_n) converges uniformly to $f = 0$ on $[0, \infty)$. However, $\int f_n \, d\lambda = -1$ whereas $\int f \, d\lambda = 0$, so

$$\liminf \int f_n \, d\lambda = -1 < 0 = \int f \, d\lambda.$$

Hence Fatou's Lemma 4.8 may not hold unless $f_n \geqslant 0$, even in the presence of uniform convergence.

4.O. Fatou's Lemma has an extension to a case where the f_n take on negative values. Let h be in $M^+(X, \boldsymbol{X})$, and suppose that $\int h \, d\mu < +\infty$. If (f_n) is a sequence in $M(X, \boldsymbol{X})$ and if $-h \leqslant f_n$, then

$$\int (\liminf f_n) \, d\mu \leqslant \liminf \int f_n \, d\mu.$$

4.P. Why doesn't Exercise 4.O apply to Exercise 4.N?

4.Q. If $f \in M^+(X, \boldsymbol{X})$ and

$$\int f \, d\mu < +\infty,$$

then $\mu\{x \in X : f(x) = +\infty\} = 0$. [*Hint:* If $E_n = \{x \in X : f(x) \geqslant n\}$, then $n\chi_{E_n} \leqslant f$.]

4.R. If $f \in M^+(X, \boldsymbol{X})$ and

$$\int f \, d\mu < +\infty,$$

then the set $N = \{x \in X : f(x) > 0\}$ is σ-finite (that is, there exists a sequence (F_n) in $\boldsymbol{X}$ such that $N \subseteq \bigcup F_n$ and $\mu(F_n) < +\infty$).

4.S. If $f \in M^+(X, \boldsymbol{X})$ and

$$\int f \, d\mu < +\infty,$$

then for any $\varepsilon > 0$ there exists a set $E \in \boldsymbol{X}$ such that $\mu(E) < +\infty$ and

$$\int f \, d\mu \leqslant \int_E f \, d\mu + \varepsilon.$$

4.T. Suppose that $(f_n) \subset M^+(X, \boldsymbol{X})$, that (f_n) converges to f, and that

$$\int f \, d\mu = \lim \int f_n \, d\mu < +\infty.$$

Prove that

$$\int_E f\,d\mu = \lim \int_E f_n\,d\mu$$

for each $E \in \boldsymbol{X}$.

4.U. Show that the conclusion of Exercise 4.T may fail if the condition

$$\lim \int f_n\,d\mu < +\infty$$

is dropped.

CHAPTER 5

Integrable Functions

In Definition 4.4 we defined the integral of each function in $M^+ = M^+(X, \boldsymbol{X})$ with respect to a measure μ and permitted this integral to be $+\infty$. In this chapter we shall discuss the integration of measurable functions which may take on both positive and negative real values. Here it is more convenient to require the values of the functions and the integral to be finite real numbers.

5.1 Definition. The collection $L = L(X, \boldsymbol{X}, \mu)$ of **integrable** (or **summable) functions** consists of all real-valued $\boldsymbol{X}$-measurable functions f defined on X, such that both the positive and negative parts f^+, f^-, of f have finite integrals with respect to μ. In this case, we define the **integral of** f **with respect to** μ to be

$$\int f\, d\mu = \int f^+\, d\mu - \int f^-\, d\mu. \tag{5.1}$$

If E belongs to $\boldsymbol{X}$, we define

$$\int_E f\, d\mu = \int_E f^+\, d\mu - \int_E f^-\, d\mu.$$

Although the integral of f is defined to be the difference of the integrals of f^+, f^-, it is easy to see that if $f = f_1 - f_2$ where f_1, f_2 are any nonnegative measurable functions with finite integrals, then

$$\int f\, d\mu = \int f_1\, d\mu - \int f_2\, d\mu.$$

In fact, since $f^+ - f^- = f = f_1 - f_2$, it follows that $f^+ + f_2 = f_1 + f^-$. If we apply Corollary 4.7(b), we infer that

$$\int f^+ \, d\mu + \int f_2 \, d\mu = \int f_1 \, d\mu + \int f^- \, d\mu.$$

Since all these terms are finite, we obtain

$$\int f \, d\mu = \int f^+ \, d\mu - \int f^- \, d\mu = \int f_1 \, d\mu - \int f_2 \, d\mu.$$

5.2 LEMMA. *If f belongs to L and λ is defined on $\boldsymbol{X}$ to $\boldsymbol{R}$ by*

$$\lambda(E) = \int_E f \, d\mu, \tag{5.2}$$

then λ is a charge.

PROOF. Since f^+ and f^- belong to M^+, Corollary 4.9 implies that the functions λ^+ and λ^-, defined by

$$\lambda^+(E) = \int_E f^+ \, d\mu, \quad \lambda^-(E) = \int_E f^- \, d\mu,$$

are measures on $\boldsymbol{X}$; they are finite because $f \in L$. Since $\lambda = \lambda^+ - \lambda^-$, it follows that λ is a charge. Q.E.D.

The function λ defined in (5.2) is frequently called the **indefinite integral of** f (with respect to μ). Since λ is a charge, if (E_n) is a disjoint sequence in $\boldsymbol{X}$ with union E, then

$$\int_E f \, d\mu = \sum_{n=1}^{\infty} \int_{E_n} f \, d\mu.$$

We refer to this relation by saying that the *indefinite integral of a function in L is countably additive.*

The next result is sometimes referred to as the *property of absolute integrability* of the Lebesgue integral. The reader will recall that, although the absolute value of a (proper) Riemann integrable function is Riemann integrable, this may no longer be the case for a function which has an improper Riemann integral (for example, consider $f(x) = x^{-1} \sin x$ on the infinite interval $1 \leqslant x < +\infty$).

5.3 THEOREM. *A measurable function f belongs to L if and only if $|f|$ belongs to L. In this case*

$$\left|\int f\,d\mu\right| \leqslant \int |f|\,d\mu. \tag{5.3}$$

PROOF. By definition f belongs to L if and only if f^+ and f^- belong to M^+ and have finite integrals. Since $|f|^+ = |f| = f^+ + f^-$ and $|f|^- = 0$, the assertion follows from Lemma 4.5(a) and Corollary 4.7(b). Moreover,

$$\begin{aligned}\left|\int f\,d\mu\right| &= \left|\int f^+\,d\mu - \int f^-\,d\mu\right| \\ &\leqslant \int f^+\,d\mu + \int f^-\,d\mu = \int |f|\,d\mu.\end{aligned}$$

Q.E.D.

5.4 COROLLARY. *If f is measurable, g is integrable, and $|f| \leqslant |g|$, then f is integrable, and*

$$\int |f|\,d\mu \leqslant \int |g|\,d\mu.$$

PROOF. This follows from Lemma 4.5(a) and Theorem 5.3. Q.E.D.

We shall now show that *the integral is linear on the space L* in the following sense.

5.5 THEOREM. *A constant multiple αf and a sum $f + g$ of functions in L belongs to L and*

$$\int \alpha f\,d\mu = \alpha \int f\,d\mu, \quad \int (f+g)\,d\mu = \int f\,d\mu + \int g\,d\mu.$$

PROOF. If $\alpha = 0$, then $\alpha f = 0$ everywhere so that

$$\int \alpha f\,d\mu = 0 = \alpha \int f\,d\mu.$$

If $\alpha > 0$, then $(\alpha f)^+ = \alpha f^+$ and $(\alpha f)^- = \alpha f^-$, whence

$$\begin{aligned}\int \alpha f\,d\mu &= \int \alpha f^+\,d\mu - \int \alpha f^-\,d\mu \\ &= \alpha\left\{\int f^+\,d\mu - \int f^-\,d\mu\right\} = \alpha \int f\,d\mu.\end{aligned}$$

The case $\alpha < 0$ is handled similarly.

If f and g belong to L, then $|f|$ and $|g|$ belong to L. Since $|f + g| \leqslant |f| + |g|$ it follows from Corollaries 4.7 and 5.4 that $f + g$ belongs to L. To establish the desired relation, we observe that

$$f + g = (f^+ + g^+) - (f^- + g^-).$$

Since $f^+ + g^+$ and $f^- + g^-$ are nonnegative integrable functions, it follows from the observation made after Definition 5.1 that

$$\int (f + g)\, d\mu = \int (f^+ + g^+)\, d\mu - \int (f^- + g^-)\, d\mu.$$

If we apply Corollary 4.7(b) and rearrange the terms, we obtain

$$\begin{aligned} \int (f + g)\, d\mu &= \int f^+\, d\mu - \int f^-\, d\mu + \int g^+\, d\mu - \int g^-\, d\mu \\ &= \int f\, d\mu + \int g\, d\mu. \end{aligned}$$

Q.E.D.

We shall now establish the most important convergence theorem for integrable functions.

5.6 Lebesgue Dominated Convergence Theorem. *Let (f_n) be a sequence of integrable functions which converges almost everywhere to a real-valued measurable function f. If there exists an integrable function g such that $|f_n| \leqslant g$ for all n, then f is integrable and*

$$\int f\, d\mu = \lim \int f_n\, d\mu. \tag{5.4}$$

PROOF. By redefining the functions f_n, f on a set of measure 0 we may assume that the convergence takes place on all of X. It follows from Corollary 5.4 that f is integrable. Since $g + f_n \geqslant 0$, we can apply Fatou's Lemma 4.8 and Theorem 5.5 to obtain

$$\begin{aligned} \int g\, d\mu + \int f\, d\mu = \int (g + f)\, d\mu &\leqslant \liminf \int (g + f_n)\, d\mu \\ &= \liminf \left(\int g\, d\mu + \int f_n\, d\mu \right) \\ &= \int g\, d\mu + \liminf \int f_n\, d\mu. \end{aligned}$$

Therefore, it follows that

$$(5.5) \qquad \int f\,d\mu \leqslant \liminf \int f_n\,d\mu.$$

Since $g - f_n \geqslant 0$, another application of Fatou's Lemma and Theorem 5.5 yields

$$\int g\,d\mu - \int f\,d\mu = \int (g - f)\,d\mu \leqslant \liminf \int (g - f_n)\,d\mu$$
$$= \int g\,d\mu - \limsup \int f_n\,d\mu,$$

from which it follows that

$$(5.6) \qquad \limsup \int f_n \leqslant \int f\,d\mu.$$

Combine (5.5) and (5.6) to infer that

$$\int f\,d\mu = \lim \int f_n\,d\mu. \qquad \text{Q.E.D.}$$

DEPENDENCE ON A PARAMETER

Frequently one needs to consider integrals where the integrand depends on a real parameter. We shall show how the Lebesgue Dominated Convergence Theorem can be used in this connection.

For the remainder of this chapter we shall let f denote a function defined on $X \times [a, b]$ to $\boldsymbol{R}$ and shall assume that the function $x \to f(x, t)$ is $\boldsymbol{X}$-measurable for each $t \in [a, b]$. Additional hypotheses will be stated explicitly.

5.7 COROLLARY. *Suppose that for some t_0 in $[a, b]$*

$$(5.7) \qquad f(x, t_0) = \lim_{t \to t_0} f(x, t)$$

for each $x \in X$, and that there exists an integrable function g on X such that $|f(x, t)| \leqslant g(x)$. Then

$$\int f(x, t_0)\,d\mu(x) = \lim_{t \to t_0} \int f(x, t)\,d\mu(x).$$

PROOF. Let (t_n) be a sequence in $[a, b]$ which converges to t_0, and apply the Dominated Convergence Theorem to the sequence (f_n) defined by $f_n(x) = f(x, t_n)$ for $x \in X$. Q.E.D.

5.8 COROLLARY. *If the function $t \to f(x, t)$ is continuous on $[a, b]$ for each $x \in X$, and if there is an integrable function g on X such that $|f(x, t)| \leqslant g(x)$, then the function F defined by*

$$F(t) = \int f(x, t)\, d\mu(x) \tag{5.8}$$

is continuous for t in $[a, b]$.

PROOF. This is an immediate consequence of Corollary 5.7. Q.E.D.

5.9 COROLLARY. *Suppose that for some $t_0 \in [a, b]$, the function $x \to f(x, t_0)$ is integrable on X, that $\partial f/\partial t$ exists on $X \times [a, b]$, and that there exists an integrable function g on X such that*

$$\left| \frac{\partial f}{\partial t}(x, t) \right| \leqslant g(x).$$

Then the function F defined in Corollary 5.8 is differentiable on $[a, b]$ and

$$\frac{dF}{dt}(t) = \frac{d}{dt} \int f(x, t)\, d\mu(x) = \int \frac{\partial f}{\partial t}(x, t)\, d\mu(x).$$

PROOF. Let t be any point of $[a, b]$. If (t_n) is a sequence in $[a, b]$ converging to t with $t_n \neq t$, then

$$\frac{\partial f}{\partial t}(x, t) = \lim \frac{f(x, t_n) - f(x, t)}{t_n - t}, \qquad x \in X.$$

Therefore, the function $x \to (\partial f/\partial t)(x, t)$ is measurable.

If $x \in X$ and $t \in [a, b]$, we can apply the Mean Value Theorem (see Reference [1], page 210) to infer the existence of a s_1 between t_0 and t such that

$$f(x, t) - f(x, t_0) = (t - t_0) \frac{\partial f}{\partial t}(x, s_1).$$

Therefore we have

$$|f(x, t)| \leqslant |f(x, t_0)| + |t - t_0|\, g(x),$$

which shows that the function $x \to f(x, t)$ is integrable for each t in $[a, b]$. Hence, if $t_n \neq t$, then

$$\frac{F(t_n) - F(t)}{t_n - t} = \int \frac{f(x, t_n) - f(x, t)}{t_n - t} \, d\mu(x).$$

Since this integrand is dominated by $g(x)$, we may apply the Dominated Convergence Theorem to obtain the stated conclusion. Q.E.D.

5.10 COROLLARY. *Under the hypotheses of Corollary 5.8,*

$$\begin{aligned} \int_a^b F(t) \, dt &= \int_a^b \left[\int f(x, t) \, d\mu(x) \right] dt \\ &= \int \left[\int_a^b f(x, t) \, dt \right] d\mu(x), \end{aligned}$$

where the integrals with respect to t are Riemann integrals.

PROOF. Recall that if φ is continuous on $[a, b]$ then

$$\frac{d}{dt} \int_a^t \varphi(s) \, ds = \varphi(t), \qquad a \leqslant t \leqslant b.$$

Let h be defined on $X \times [a, b]$ by

$$h(x, t) = \int_a^t f(x, s) \, ds.$$

It follows that $(\partial h / \partial t)(x, t) = f(x, t)$. Since this Riemann integral exists, it is the limit of a sequence of Riemann sums; hence the map $x \to h(x, t)$ is measurable for each t. Moreover, since $|f(x, t)| \leqslant g(x)$, we infer that $|h(x, t)| \leqslant g(x)(b - a)$, so that the function $x \to h(x, t)$ is integrable for each $t \in [a, b]$. Let H be defined on $[a, b]$ by

$$H(t) = \int h(x, t) \, d\mu(x);$$

it follows from Corollary 5.9 that

$$\frac{dH}{dt}(t) = \int \frac{\partial h}{\partial t}(x, t) \, d\mu(x) = \int f(x, t) \, d\mu(x) = F(t).$$

Therefore we have

$$\begin{aligned}\int_a^b F(t)\,dt &= H(b) - H(a)\\ &= \int [h(x,b) - h(x,a)]\,d\mu(x)\\ &= \int \left[\int_a^b f(x,t)\,dt\right] d\mu(x).\end{aligned}$$

Q.E.D.

The interchange of the order of (Lebesgue) integrals will be considered in Chapter 10.

EXERCISES

5.A. If $f \in L(X, \boldsymbol{X}, \mu)$ and $a > 0$, show that the set $\{x \in X : |f(x)| \geqslant a\}$ has finite measure. In addition, the set $\{x \in X : f(x) \neq 0\}$ has σ-finite measure (that is, the union of a sequence of measurable sets with finite measure).

5.B. If f is an $\boldsymbol{X}$-measurable real-valued function and if $f(x) = 0$ for μ-almost all x in X, then $f \in L(X, \boldsymbol{X}, \mu)$ and

$$\int f\,d\mu = 0.$$

5.C. If $f \in L(X, \boldsymbol{X}, \mu)$ and g is an $\boldsymbol{X}$-measurable real-valued function such that $f(x) = g(x)$ almost everywhere on X, then $g \in L(X, \boldsymbol{X}, \mu)$ and

$$\int f\,d\mu = \int g\,d\mu.$$

5.D. If $f \in L(X, \boldsymbol{X}, \mu)$ and $\varepsilon > 0$, then there exists a measurable simple function φ such that

$$\int |f - \varphi|\,d\mu < \varepsilon.$$

5.E. If $f \in L$ and g is a bounded measurable function, then the product fg also belongs to L.

5.F. If f belongs to L, then it does not follow that f^2 belongs to L.

5.G. Suppose that f is in $L(X, \mathbf{X}, \mu)$ and that its indefinite integral is

$$\lambda(E) = \int_E f \, d\mu, \qquad E \in \mathbf{X}.$$

Show that $\lambda(E) \geqslant 0$ for all $E \in \mathbf{X}$ if and only if $f(x) \geqslant 0$ for almost all $x \in X$. Moreover, $\lambda(E) = 0$ for all E if and only if $f(x) = 0$ for almost all $x \in X$.

5.H. Suppose that f_1 and f_2 are in $L(X, \mathbf{X}, \mu)$ and let λ_1 and λ_2 be their indefinite integrals. Show that $\lambda_1(E) = \lambda_2(E)$ for all $E \in \mathbf{X}$ if and only if $f_1(x) = f_2(x)$ for almost all x in X.

5.I. If f is a complex-valued function on X such that $\operatorname{Re} f$ and $\operatorname{Im} f$ belong to $L(X, \mathbf{X}, \mu)$, we say that f is **integrable** and define

$$\int f \, d\mu = \int \operatorname{Re} f \, d\mu + i \int \operatorname{Im} f \, d\mu.$$

Let f be a complex-valued measurable function. Show that f is integrable if and only if $|f|$ is integrable, in which case

$$\left| \int f \, d\mu \right| \leqslant \int |f| \, d\mu.$$

[*Hint:* If $\int f \, d\mu = r \, e^{i\theta}$ with r, θ real, consider $g(x) = e^{-i\theta} f(x)$.]

5.J. Let (f_n) be a sequence of complex-valued measurable functions which converges to f. If there exists an integrable function g such that $|f_n| \leqslant g$, show that

$$\int f \, d\mu = \lim \int f_n \, d\mu.$$

5.K. Let $X = \mathbf{N}$, let $\mathbf{X}$ be all subsets of $\mathbf{N}$, and let μ be the counting measure on $\mathbf{X}$. Show that f belongs to $L(X, \mathbf{X}, \mu)$ if and only if the series $\sum f(n)$ is absolutely convergent, in which case

$$\int f \, d\mu = \sum_{n=1}^{\infty} f(n).$$

5.L. If (f_n) is a sequence in $L(X, \mathbf{X}, \mu)$ which converges uniformly on X to a function f, and if $\mu(X) < +\infty$, then

$$\int f \, d\mu = \lim \int f_n \, d\mu.$$

5.M. Show that the conclusion in the Exercise 5.L may fail if the hypothesis $\mu(X) < +\infty$ is dropped.

5.N. Let $f_n = n\,\chi_{[0,1/n]}$, where $X = \boldsymbol{R}$, $\boldsymbol{X} = \boldsymbol{B}$, and μ is a Lebesgue measure. Show that the condition $|f_n| \leqslant g$ cannot be dropped in the Lebesgue Dominated Convergence Theorem.

5.O. If $f_n \in L(X, \boldsymbol{X}, \mu)$, and if

$$\sum_{n=1}^{\infty} \int |f_n| \, d\mu < +\infty,$$

then the series $\sum f_n(x)$ converges almost everywhere to a function f in $L(X, \boldsymbol{X}, \mu)$. Moreover,

$$\int f \, d\mu = \sum_{n=1}^{\infty} \int f_n \, d\mu.$$

5.P. Let $f_n \in L(X, \boldsymbol{X}, \mu)$, and suppose that (f_n) converges to a function f. Show that if

$$\lim \int |f_n - f| \, d\mu = 0, \qquad \text{then} \quad \int |f| \, d\mu = \lim \int |f_n| \, d\mu.$$

5.Q. If $t > 0$, then

$$\int_0^{+\infty} e^{-tx} \, dx = \frac{1}{t}.$$

Moreover, if $t \geqslant a > 0$, then $e^{-tx} \leqslant e^{-ax}$. Use this and Exercise 4.M to justify differentiating under the integral sign and to obtain the formula

$$\int_0^{+\infty} x^n e^{-x} \, dx = n!$$

5.R. Suppose that f is defined on $X \times [a, b]$ to $\boldsymbol{R}$ and that the function $x \to f(t, x)$ is $\boldsymbol{X}$-measurable for each $t \in [a, b]$. Suppose that for some t_0 and t_1 in $[a, b]$ the function $x \to f(x, t_0)$ is integrable on X, that $(\partial f/\partial t)(x, t_1)$ exists, and that there exists an integrable function g on X such that

$$\left| \frac{f(x, t) - f(x, t_1)}{t - t_1} \right| \leqslant g(x)$$

for $x \in X$, and $t \in [a, b]$, $t \neq t_1$. Then

$$\left[\frac{d}{dt} \int f(x, t) \, d\mu(x) \right]_{t=t_1} = \int \frac{\partial f}{\partial t}(x, t_1) \, d\mu(x).$$

5.S. Suppose the function $x \to f(x, t)$ is $\boldsymbol{X}$-measurable for each $t \in \boldsymbol{R}$, and the function $t \to f(x, t)$ is continuous on $\boldsymbol{R}$ for each $x \in X$. In addition, suppose that there are integrable functions g, h on X such that $|f(x, t)| \leqslant g(x)$ and such that the improper Riemann integral

$$\int_{-\infty}^{+\infty} |f(x, t)| \, dt \leqslant h(x).$$

Show that

$$\int_{-\infty}^{+\infty} \left[\int f(x, t) \, d\mu(x) \right] dt = \int \left[\int_{-\infty}^{+\infty} f(x, t) \, dt \right] d\mu(x),$$

where the integrals with respect to t are improper Riemann integrals.

5.T. Let f be an $\boldsymbol{X}$-measurable function on X to $\boldsymbol{R}$. For $n \in \boldsymbol{N}$, let (f_n) be the sequence of truncates of f (see Exercise 2.K). If f is integrable with respect to μ, then

$$\int f \, d\mu = \lim \int f_n \, d\mu.$$

Conversely, if

$$\sup \int |f_n| \, d\mu < +\infty,$$

then f is integrable.

CHAPTER 6

The Lebesgue Spaces L_p

It is often useful to impose the structure of a Banach space on the set of all integrable functions on a measure space $(X, \boldsymbol{X}, \mu)$. In addition, we shall introduce the L_p, $1 \leqslant p \leqslant \infty$, spaces which occur frequently in analysis. Aside from the intrinsic importance of these spaces, we examine them here partly to indicate applications of some of the results in the earlier sections.

6.1 DEFINITION. If V is a real linear (= vector) space, then a real-valued function N on V is said to be a **norm** for V in case it satisfies

(i) $N(v) \geqslant 0$ for all $v \in V$;
(ii) $N(v) = 0$ if and only if $v = 0$;
(iii) $N(\alpha v) = |\alpha| N(v)$ for all $v \in V$ and real α;
(iv) $N(u + v) \leqslant N(u) + N(v)$ for all $u, v \in V$.

If condition (ii) is dropped, the function N is said to be a **semi-norm** or a **pseudo-norm** for V. A **normed linear space** is a linear space V together with a norm for V.

6.2 EXAMPLES. (a) The absolute value function yields a norm for the real numbers.

(b) The linear space $\boldsymbol{R}^n$ of n-tuples of real numbers can be normed by defining

$$N_1(u_1, \ldots, u_n) = |u_1| + \cdots + |u_n|,$$
$$N_p(u_1, \ldots, u_n) = \{|u_1|^p + \cdots + |u_n|^p\}^{1/p}, \quad p \geqslant 1,$$
$$N_\infty(u_1, \ldots, u_n) = \sup\{|u_1|, \ldots, |u_n|\}.$$

It is easy to check that N_1 and N_∞ are norms and that N_p satisfies (i), (ii), (iii). It is a consequence of Minkowski's Inequality, which will be proved subsequently, that N_p satisfies (iv).

(c) The linear space l_1 of all real-valued sequences $u = (u_n)$ such that $N_1(u) = \sum |u_n| < +\infty$ is a normed linear space under N_1. Similarly, if $1 \leqslant p < \infty$, the collection l_p of all sequences such that $N_p(u) = \{\sum |u_n|^p\}^{1/p} < +\infty$ is normed by N_p.

(d) The collection of all real-valued functions defined on an infinite set X cannot be normed, but the collection $B(X)$ of all bounded real-valued functions on X is normed by

$$N(f) = \sup \{|f(x)| : x \in X\}.$$

In particular, the linear space of continuous functions on $X = [a, b]$ is normed.

All the preceding examples have been proper norms on a linear space. There are also semi-norms on a linear space that are of interest. The following are some examples.

6.3 Examples. (a) On the space $\boldsymbol{R}^n$, consider the semi-norm

$$N_0(u_1, \ldots, u_n) = \sup \{|u_2|, \ldots, |u_n|\}.$$

Here $N_0(u_1, \ldots, u_n) = 0$ if and only if $u_2 = \cdots = u_n = 0$.

(b) On the linear space $C[0, 1]$ of continuous functions on $[0, 1]$ to $\boldsymbol{R}$, define the semi-norm

$$N_0(f) = \sup \{|f(x)| : 0 \leqslant x \leqslant \tfrac{1}{2}\}.$$

Here $N_0(f) = 0$ if and only if $f(x)$ vanishes for $0 \leqslant x \leqslant \frac{1}{2}$.

(c) On the linear space of functions on $[a, b]$ to $\boldsymbol{R}$ which have continuous derivatives, consider the semi-norm

$$N_0(f) = \sup \{|f'(x)| : a \leqslant x \leqslant b\}.$$

Here $N_0(f) = 0$ if and only if f is constant on $[a, b]$.

6.4 Definition. Let $(X, \boldsymbol{X}, \mu)$ be a measure space. If f belongs to $L(X, \boldsymbol{X}, \mu)$, we define

$$N_\mu(f) = \int |f| \, d\mu.$$

It will be shown that N_μ is a semi-norm on the space $L(X, \boldsymbol{X}, \mu)$.

6.5 LEMMA. *The space $L(X, \boldsymbol{X}, \mu)$ is a linear space under the operations defined by*

$$(f + g)(x) = f(x) + g(x), \qquad (\alpha f)(x) = \alpha f(x), \quad x \in X,$$

and N_μ is a semi-norm on $L(X, \boldsymbol{X}, \mu)$. Moreover, $N_\mu(f) = 0$ if and only if $f(x) = 0$ for μ-almost all x in X.

PROOF. It was seen in Theorem 5.5 that $L = L(X, \boldsymbol{X}, \mu)$ is a linear space under the indicated operations. It is clear that $N_\mu(f) \geqslant 0$ for $f \in L$, and that

$$N_\mu(\alpha f) = \int |\alpha f| \, d\mu = |\alpha| \int |f| \, d\mu = |\alpha| N_\mu(f).$$

Moreover, it follows from the Triangle Inequality that

$$\begin{aligned} N_\mu(f + g) &= \int |f + g| \, d\mu \leqslant \int (|f| + |g|) \, d\mu \\ &= \int |f| \, d\mu + \int |g| \, d\mu = N_\mu(f) + N_\mu(g). \end{aligned}$$

Hence N_μ is a semi-norm on L, and it follows from Corollary 4.10 that $N_\mu(f) = 0$ if and only if $f(x) = 0$ for almost all x. Q.E.D.

In order to make $L(X, \boldsymbol{X}, \mu)$ into a normed linear space, we shall identify two functions that are equal almost everywhere; that is, we use equivalence classes of functions instead of functions.

6.6 DEFINITION. Two functions in $L = L(X, \boldsymbol{X}, \mu)$ are said to be **μ-equivalent** if they are equal μ-almost everywhere. The **equivalence class determined by** f in L is sometimes denoted by $[f]$ and consists of the set of all functions in L which are μ-equivalent to f. The **Lebesgue space** $L_1 = L_1(X, \boldsymbol{X}, \mu)$ consists of all μ-equivalence classes in L. If $[f]$ belongs to L_1, we define its **norm** by

$$\|[f]\|_1 = \int |f| \, d\mu. \tag{6.1}$$

6.7 THEOREM. *The Lebesgue space $L_1(X, \boldsymbol{X}, \mu)$ is a normed linear space.*

PROOF. It is understood, of course, that the vector operations in L_1 are defined by

$$\alpha[f] = [\alpha f], \quad [f] + [g] = [f + g],$$

and that the zero element of L_1 is $[0]$. We shall check only that equation (6.1) gives a norm on L_1. Certainly $\|[f]\|_1 \geqslant 0$ and $\|[0]\|_1 = 0$. Moreover, if $\|[f]\|_1 = 0$ then

$$\int |f| \, d\mu = 0,$$

so $f(x) = 0$ for μ-almost all x. Hence $[f] = [0]$. Finally, it is easily seen that properties (iii) and (iv) of Definition 6.1 are satisfied. Therefore $\| \;\|_1$ yields a norm on L_1. Q.E.D.

It should always be remembered that the elements of L_1 are actually equivalence classes of functions in L. However, it is both convenient and customary to regard these elements as being functions, and we shall subsequently do so. Thus we shall make reference to the equivalence class $[f]$ by referring to "the element f of L_1," and we shall write $\|f\|_1$ in place of $\|[f]\|_1$.

THE SPACES L_p, $1 \leqslant p < +\infty$

We now wish to consider a family of related normed linear spaces of equivalence classes of measurable functions.

6.8 DEFINITION. If $1 \leqslant p < \infty$, the space $L_p = L_p(X, \mathbf{X}, \mu)$ consists of all μ-equivalence classes of $\mathbf{X}$-measurable real-valued functions f for which $|f|^p$ has finite integral with respect to μ over X. Two functions are **μ-equivalent** if they are equal μ-almost everywhere. We set

$$\|f\|_p = \left\{ \int |f|^p \, d\mu \right\}^{1/p}. \tag{6.3}$$

If $p = 1$, this reduces to the norm introduced previously on the space L_1 of equivalence classes of integrable functions. We shall show subsequently that if $1 \leqslant p < \infty$, then L_p is a normed linear space under (6.3), and is complete under this norm; thus L_p is a Banach space. It is understood that the vector operations between the equivalence classes in L_p are defined pointwise: the sum of the equivalence classes containing f and g is the equivalence class containing $f + g$ and similarly for the product cf.

In the special case where μ is the counting measure on all subsets of N, the L_p-spaces can be identified with the sequence spaces l_p of Example 6.2(c). In this case, each equivalence class contains one element. It is frequently enlightening to interpret assertions about general L_p-spaces by considering the somewhat simpler l_p-spaces.

In order to establish that (6.3) yields a norm on L_p, we shall need the following basic inequality.

6.9 HÖLDER'S INEQUALITY. *Let $f \in L_p$ and $g \in L_q$ where $p > 1$ and $(1/p) + (1/q) = 1$. Then $fg \in L_1$ and $\|fg\|_1 \leqslant \|f\|_p \|g\|_q$.*

PROOF. Let α be a real number satisfying $0 < \alpha < 1$, and consider the function φ defined for $t \geqslant 0$ by

$$\varphi(t) = \alpha t - t^\alpha.$$

It is easy to check that $\varphi'(t) < 0$ for $0 < t < 1$ and $\varphi'(t) > 0$ for $t > 1$. It follows from the Mean Value Theorem of calculus that $\varphi(t) \geqslant \varphi(1)$ and that $\varphi(t) = \varphi(1)$, if and only if $t = 1$. Therefore we have

$$t^\alpha \leqslant \alpha t + (1 - \alpha), \qquad t \geqslant 0.$$

If a, b are nonnegative, and if we let $t = a/b$ and multiply by b, we obtain the inequality

$$a^\alpha b^{1-\alpha} \leqslant \alpha a + (1 - \alpha)b,$$

where equality holds if and only if $a = b$.

Now let p and q satisfy $1 < p < \infty$ and $(1/p) + (1/q) = 1$ and take $\alpha = 1/p$. It follows that if A, B are any nonnegative real numbers, then

$$AB \leqslant \frac{A^p}{p} + \frac{B^q}{q}. \tag{6.4}$$

and that the equality holds if and only if $A^p = B^q$.

Suppose that $f \in L_p$ and $g \in L_q$, and that $\|f\|_p \neq 0$ and $\|g\|_q \neq 0$.

The product fg is measurable and (6.4) with $A = |f(x)|/\|f\|_p$ and $B = |g(x)|/\|g\|_q$ implies that

$$\frac{|f(x)g(x)|}{\|f\|_p\ \|g\|_q} \leqslant \frac{|f(x)|^p}{p\|f\|_p{}^p} + \frac{|g(x)|^q}{q\|g\|_q{}^q}.$$

Since both of the terms on the right are integrable, it follows from Corollary 5.4 and Theorem 5.5 that fg is integrable. Moreover, on integrating we obtain

$$\frac{\|fg\|_1}{\|f\|_p\|g\|_q} \leqslant \frac{1}{p} + \frac{1}{q} = 1.$$

which is Hölder's Inequality. Q.E.D.

Hölder's Inequality implies that the product of a function in L_p and a function in L_q is integrable when $p > 1$ and q satisfies the relation $(1/p) + (1/q) = 1$ or, equivalently, when $p + q = pq$. Two numbers satisfying this relation are said to be **conjugate indices**. It will be noted that $p = 2$ is the only **self-conjugate** index. Thus the product of two functions in L_2 is integrable.

6.10 CAUCHY–BUNYAKOVSKIĬ–SCHWARZ INEQUALITY. *If f and g belong to L_2, then fg is integrable and*

$$\left|\int fg\, d\mu\right| \leqslant \int |fg|\, d\mu \leqslant \|f\|_2\ \|g\|_2. \tag{6.5}$$

6.11 MINKOWSKI'S INEQUALITY. *If f and h belong to L_p, $p \geqslant 1$, then $f + h$ belongs to L_p and*

$$\|f + h\|_p \leqslant \|f\|_p + \|h\|_p. \tag{6.6}$$

PROOF. The case $p = 1$ has already been treated, so we suppose $p > 1$. The sum $f + h$ is evidently measurable. Since

$$|f + h|^p \leqslant [2 \sup\{|f|,|h|\}]^p \leqslant 2^p\{|f|^p + |h|^p\}$$

it follows from Corollary 5.4 and Theorem 5.5 that $f + h \in L_p$. Moreover,

$$|f + h|^p = |f + h|\ |f + h|^{p-1} \leqslant |f|\ |f + h|^{p-1} + |h|\ |f + h|^{p-1}. \tag{6.7}$$

Since $f + h \in L_p$, then $|f + h|^p \in L_1$; since $p = (p - 1)q$ it follows that

$|f+h|^{p-1} \in L_q$. Hence we can apply Hölder's Inequality to infer that

$$\int |f|\, |f+h|^{p-1}\, d\mu \leqslant \|f\|_p \left\{\int |f+h|^{(p-1)q}\, d\mu\right\}^{1/q}$$
$$= \|f\|_p\, \|f+h\|_p^{\,p/q}.$$

If we treat the second term on the right in (6.7) similarly, we obtain

$$\|f+h\|_p^{\,p} \leqslant \|f\|_p\, \|f+h\|_p^{\,p/q} + \|h\|_p\, \|f+h\|_p^{\,p/q}$$
$$= \{\|f\|_p + \|h\|_p\}\, \|f+h\|_p^{\,p/q}.$$

If $A = \|f+h\|_p = 0$, then equation (6.6) is trivial. If $A \neq 0$, we can divide the above inequality by $A^{p/q}$; since $p - p/q = 1$, we obtain Minkowski's Inequality. Q.E.D.

It is readily seen that the space L_p is a linear space and that formula (6.3) defines a norm on L_p. The only nontrivial thing to be checked here is the inequality 6.1(iv) and this is Minkowski's Inequality. We shall now show that L_p is complete under this norm in the following sense.

6.12 DEFINITION. A sequence (f_n) in L_p is a **Cauchy sequence** in L_p if for every positive number ε there exists an $M(\varepsilon)$ such that if $m, n \geqslant M(\varepsilon)$, then $\|f_m - f_n\|_p < \varepsilon$. A sequence (f_n) in L_p is **convergent to** f in L_p if for every positive number ε there exists an $N(\varepsilon)$ such that if $n \geqslant N(\varepsilon)$, then $\|f - f_n\|_p < \varepsilon$. A normed linear space is **complete** if every Cauchy sequence converges to some element of the space.

6.13 LEMMA. *If the sequence (f_n) converges to f in L_p, then it is a Cauchy sequence.*

PROOF. If $m, n \geqslant N(\varepsilon/2)$, then

$$\|f - f_m\|_p < \frac{\varepsilon}{2}, \qquad \|f - f_n\|_p < \frac{\varepsilon}{2}.$$

Hence we have

$$\|f_m - f_n\|_p \leqslant \|f_m - f\|_p + \|f - f_n\|_p < \varepsilon. \qquad \text{Q.E.D.}$$

We shall now show that every Cauchy sequence in L_p converges in L_p to an element. This result is sometimes called the Riesz–Fischer Theorem.

6.14 COMPLETENESS THEOREM. *If* $1 \leqslant p < \infty$, *then the space* L_p *is a complete normed linear space under the norm*

$$\|f\|_p = \left\{ \int |f|^p \, d\mu \right\}^{1/p}.$$

PROOF. It has been stated that L_p is a normed linear space. To establish the completeness of L_p, let (f_n) be a Cauchy sequence relative to the norm $\| \; \|_p$. Hence, if $\varepsilon > 0$ there exists an $M(\varepsilon)$ such that if $m, n \geqslant M(\varepsilon)$, then

$$\int |f_m - f_n|^p \, d\mu = \|f_m - f_n\|_p{}^p < \varepsilon^p. \tag{6.8}$$

There exists a subsequence (g_k) of (f_n) such that $\|g_{k+1} - g_k\|_p < 2^{-k}$ for $k \in \boldsymbol{N}$. Define g by

$$g(x) = |g_1(x)| + \sum_{k=1}^{\infty} |g_{k+1}(x) - g_k(x)|, \tag{6.9}$$

so that g is in $M^+(X, \boldsymbol{X})$. By Fatou's Lemma, we have

$$\int |g|^p \, d\mu \leqslant \liminf_{n \to \infty} \int \left\{ |g_1| + \sum_{k=1}^{n} |g_{k+1} - g_k| \right\}^p d\mu.$$

Take the pth root of both sides and apply Minkowski's Inequality to obtain

$$\begin{aligned} \left\{ \int |g|^p \, d\mu \right\}^{1/p} &\leqslant \liminf_{n \to \infty} \left\{ \|g_1\|_p + \sum_{k=1}^{n} \|g_{k+1} - g_k\|_p \right\} \\ &\leqslant \|g_1\|_p + 1. \end{aligned}$$

Hence, if $E = \{x \in X : g(x) < +\infty\}$, then $E \in \boldsymbol{X}$ and $\mu(X \setminus E) = 0$. Therefore, the series in (6.9) converges almost everywhere and $g\,\chi_E$ belongs to L_p.

We now define f on X by

$$\begin{aligned} f(x) &= g_1(x) + \sum_{k=1}^{\infty} \{g_{k+1}(x) - g_k(x)\}, \qquad x \in E, \\ &= 0, \qquad x \notin E. \end{aligned}$$

Since $|g_k| \leqslant \sum_{j=k}^{\infty} |g_{j+1} - g_j| \leqslant g$ and the sequence (g_k) converges almost everywhere to f, the Dominated Convergence Theorem 5.6

implies that $f \in L_p$. Since $|f - g_k|^p \leqslant 2^p g^p$, we infer from the Dominated Convergence Theorem that $0 = \lim \|f - g_k\|_p$, so that (g_k) converges in L_p to f.

In view of (6.8), if $m \geqslant M(\varepsilon)$ and k is sufficiently large, then

$$\int |f_m - g_k|^p \, d\mu < \varepsilon^p.$$

Apply Fatou's Lemma to conclude that

$$\int |f_m - f|^p \, d\mu \leqslant \liminf_{k \to \infty} \int |f_m - g_k|^p \, d\mu \leqslant \varepsilon^p,$$

whenever $m \geqslant M(\varepsilon)$. This proves that the sequence (f_n) converges to f in the norm of L_p. Q.E.D.

A complete normed linear space is usually called a **Banach space** Thus the preceding theorem could be formulated: *the space* L_p *is a Banach space under the norm given in* (6.3).

THE SPACE L_∞

We shall now introduce a space which is related to the L_p-spaces.

6.15 Definition. The space $L_\infty = L_\infty(X, \mathbf{X}, \mu)$ consists of all the equivalence classes of $\mathbf{X}$-measurable real-valued functions which are almost everywhere bounded, two functions being equivalent when they are equal μ-almost everywhere. If $f \in L_\infty$ and $N \in \mathbf{X}$ with $\mu(N) = 0$, we define

$$S(N) = \sup \{|f(x)| : x \notin N\}$$

and

$$\|f\|_\infty = \inf \{S(N) : N \in \mathbf{X}, \mu(N) = 0\}. \tag{6.10}$$

An element of L_∞ is called an **essentially bounded function**.

It follows (see Exercise 6.T) that if $f \in L_\infty$, then $|f(x)| \leqslant \|f\|_\infty$ for almost all x. Moreover, if $A < \|f\|_\infty$, then there exists a set E with positive measure such that $|f(x)| \geqslant A$ for $x \in E$, It is also clear that the norm in (6.10) is well-defined on L_∞.

6.16 THEOREM. *The space L_∞ is a complete normed linear space under the norm given by formula* (6.10).

PROOF. It is clear that L_∞ is a linear space and that $\|f\|_\infty \geqslant 0$, $\|0\|_\infty = 0$, and $\|\alpha f\|_\infty = |\alpha| \, \|f\|_\infty$. If $\|f\|_\infty = 0$, then there exists a set $N_k \in \boldsymbol{X}$ with $\mu(N_k) = 0$ such that $|f(x)| \leqslant 1/k$ for $x \in N_k$. If we put $N = \bigcup_{k=1}^{\infty} N_k$, then $N \in \boldsymbol{X}$, $\mu(N) = 0$, and $|f(x)| = 0$ for $x \notin N$. Therefore, $f(x) = 0$ for almost all x.

If $f, g \in L_\infty$, there exist sets N_1, N_2 in $\boldsymbol{X}$ with $\mu(N_1) = \mu(N_2) = 0$ such that

$$
\begin{aligned}
|f(x)| &\leqslant \|f\|_\infty \quad \text{for} \quad x \notin N_1, \\
|g(x)| &\leqslant \|g\|_\infty \quad \text{for} \quad x \notin N_2.
\end{aligned}
$$

Therefore $|f(x) + g(x)| \leqslant \|f\|_\infty + \|g\|_\infty$ for $x \notin (N_1 \cup N_2)$, from which it follows that $\|f + g\|_\infty \leqslant \|f\|_\infty + \|g\|_\infty$.

It remains to prove that L_∞ is complete. Let (f_n) be a Cauchy sequence in L_∞, and let M be a set in $\boldsymbol{X}$ with $\mu(M) = 0$, such that $|f_n(x)| \leqslant \|f_n\|_\infty$ for $x \notin M$, $n = 1, 2, \ldots$, and also such that $|f_n(x) - f_m(x)| \leqslant \|f_n - f_m\|_\infty$ for all $x \notin M$, n, $m = 1, 2, \ldots$. Then the sequence (f_n) is uniformly convergent on $X \setminus M$, and we let

$$
\begin{aligned}
f(x) &= \lim f_n(x), \qquad x \notin M, \\
&= 0, \qquad\qquad\quad\; x \in M.
\end{aligned}
$$

It follows that f is measurable, and it is easily seen that $\|f_n - f\|_\infty \to 0$. Hence L_∞ is complete. Q.E.D.

EXERCISES

6.A. Let $C[0, 1]$ be the linear space of continuous functions on $[0, 1]$ to $\boldsymbol{R}$. Define N_0 for f in $C[0, 1]$ by $N_0(f) = |f(0)|$. Show that N_0 is a semi-norm on $C[0, 1]$.

6.B. Let $C[0, 1]$ be as before and define N_1 for f in $C[0, 1]$ to be the Riemann integral of $|f|$ over $[0, 1]$. Show that N_1 is a semi-norm on $C[0, 1]$. If f_n is defined for $n \geqslant 1$ to be equal to 0 for $0 \leqslant x \leqslant (1 - 1/n)/2$, to be equal to 1 for $\frac{1}{2} \leqslant x \leqslant 1$, and to be linear for $(1 - 1/n)/2 \leqslant x \leqslant \frac{1}{2}$, show that (f_n) is a Cauchy sequence, but that it does not converge relative to N_1 to an element of $C[0, 1]$.

6.C. Let N be a norm on a linear space V and let d be defined for $u, v \in V$ by $d(u, v) = N(u - v)$. Show that d is a metric on N; that is, (i) $d(u, v) \geqslant 0$ for all $u, v \in V$; (ii) $d(u, v) = 0$ if and only if $u = v$; (iii) $d(u, v) = d(v, u)$; (iv) $d(u, v) \leqslant d(u, w) + d(w, v)$.

6.D. If $f \in L_1(X, \boldsymbol{X}, \mu)$ and $\varepsilon > 0$, then there exists a simple $\boldsymbol{X}$-measurable function φ such that $\|f - \varphi\|_1 < \varepsilon$. Extend this to L_p, $1 \leqslant p < \infty$. Is this true for L_∞?

6.E. If $f \in L_p$, $1 \leqslant p < \infty$, and if $E = \{x \in X : |f(x)| \neq 0\}$, then E is σ-finite.

6.F. If $f \in L_p$ and if $E_n = \{x \in X : |f(x)| \geqslant n\}$, then $\mu(E_n) \to 0$ as $n \to \infty$.

6.G. Let $X = \boldsymbol{N}$, and let μ be the counting measure on $\boldsymbol{N}$. If f is defined on $\boldsymbol{N}$ by $f(n) = 1/n$, then f does not belong to L_1, but it does belong to L_p for $1 < p \leqslant \infty$. [Alternatively, let $X = \boldsymbol{R}$, $\boldsymbol{X} = \boldsymbol{B}$, and let μ be Lebesgue measure and define $g(x) = 0$ for $x < 1$ and $g(x) = 1/x$ for $x \geqslant 1$.]

6.H. Let $X = \boldsymbol{N}$, and let λ be the measure on $\boldsymbol{N}$ which has measure $1/n^2$ at the point n. (More precisely $\lambda(E) = \sum\{1/n^2 : n \in E\}$.) Show that $\lambda(X) < +\infty$. Let f be defined on X by $f(n) = \sqrt{n}$. Show that $f \in L_p$ if and only if $1 \leqslant p < 2$. [For a similar example, let $X = (0, 1)$ with Lebesgue measure, and consider $g(x) = 1/\sqrt{x}$.]

6.I. Modify the Exercise 6.H to obtain a function on a finite measure space which belongs to L_p if and only if $1 \leqslant p \leqslant p_0$.

6.J. Let $(X, \boldsymbol{X}, \mu)$ be a finite measure space. If f is $\boldsymbol{X}$-measurable, let $E_n = \{x \in X : (n - 1) \leqslant |f(x)| < n\}$. Show that $f \in L_1$ if and only if

$$\sum_{n=1}^{\infty} n\,\mu(E_n) < +\infty.$$

More generally, $f \in L_p$ for $1 \leqslant p < \infty$, if and only if

$$\sum_{n=1}^{\infty} n^p\,\mu(E_n) < +\infty.$$

6.K. If $(X, \boldsymbol{X}, \mu)$ is a finite measure space and $f \in L_p$, then $f \in L_r$ for $1 \leqslant r \leqslant p$. (*Hint:* Use Exercise 6.J or the inequality $|f|^r \leqslant 1 +$

$|f|^p$.) Apply Hölder's Inequality to $|f|^r$ in $L_{p/r}$ and $g = 1$ to obtain the inequality

$$\|f\|_r \leqslant \|f\|_p \, \mu(X)^s,$$

where $s = (1/r) - (1/p)$. Therefore, if $\mu(X) = 1$, then $\|f\|_r \leqslant \|f\|_p$.

6.L. Suppose that $X = \boldsymbol{N}$ and μ is the counting measure on $\boldsymbol{N}$. If $f \in L_p$, then $f \in L_s$ with $1 \leqslant p \leqslant s < \infty$, and $\|f\|_s \leqslant \|f\|_p$.

6.M. Let $X = (0, \infty)$, let μ be Lebesgue measure on X, and let $f(x) = x^{-1/2}(1 + |\log x|)^{-1}$. Then $f \in L_p$ if and only if $p = 2$.

6.N. Let $(X, \boldsymbol{X}, \mu)$ be any measure space and let f belong to both L_{p_1} and L_{p_2}, with $1 \leqslant p_1 < p_2 < \infty$. Prove that $f \in L_p$ for any value of p such that $p_1 \leqslant p \leqslant p_2$.

6.O. Let $1 < p < \infty$, and let $(1/p) + (1/q) = 1$. It follows from Hölder's Inequality that if $f \in L_p$, then

$$\left| \int fg \, d\mu \right| \leqslant \|f\|_p$$

for all $g \in L_q$ such that $\|g\|_q \leqslant 1$. If $f \neq 0$, define g_0 on X by $g_0(x) = c[\operatorname{signum} f(x)] f(x)^{p-1}$, where $c = (\|f\|_p)^{-p/q}$. Show that $g_0 \in L_q$, that $\|g_0\|_q = 1$, and that

$$\left| \int f g_0 \, d\mu \right| = \|f\|_p.$$

6.P. Let $f \in L_p(X, \boldsymbol{X}, \mu)$, $1 \leqslant p < \infty$, and let $\varepsilon > 0$. Show that there exists a set $E_\varepsilon \in \boldsymbol{X}$ with $\mu(E_\varepsilon) < +\infty$ such that if $F \in \boldsymbol{X}$ and $F \cap E_\varepsilon = \emptyset$, then $\|f \chi_F\|_p < \varepsilon$.

6.Q. Let $f_n \in L_p(X, \boldsymbol{X}, \mu)$, $1 \leqslant p < \infty$, and let β_n be defined for $E \in \boldsymbol{X}$ by

$$\beta_n(E) = \left\{ \int_E |f_n|^p \, d\mu \right\}^{1/p}.$$

Show that $|\beta_n(E) - \beta_m(E)| \leqslant \|f_n - f_m\|_p$. Hence, if (f_n) is a Cauchy sequence in L_p, then $\lim \beta_n(E)$ exists for each $E \in \boldsymbol{X}$.

6.R. Let f_n, β_n be as in Exercise 6.Q. If (f_n) is a Cauchy sequence and $\varepsilon > 0$, then there exists a set $E_\varepsilon \in \boldsymbol{X}$ with $\mu(E_\varepsilon) < +\infty$ such that if $F \in \boldsymbol{X}$ and $F \cap E_\varepsilon = \emptyset$, then $\beta_n(F) < \varepsilon$ for all $n \in \boldsymbol{N}$.

6.S. Let f_n, β_n be as in the Exercise 6.R, and suppose that (f_n) is a Cauchy sequence. If $\varepsilon > 0$, then there exists a $\delta(\varepsilon) > 0$ such that if $E \in \boldsymbol{X}$ and $\mu(E) < \delta(\varepsilon)$, then $\beta_n(E) < \varepsilon$ for all $n \in \boldsymbol{N}$. (*Hint:* Use Corollary 4.11.)

6.T. If $f \in L_\infty(X, \boldsymbol{X}, \mu)$, then $|f(x)| \leqslant \|f\|_\infty$ for almost all x. Moreover, if $A < \|f\|_\infty$, then there exists a set $E \in \boldsymbol{X}$ with $\mu(E) > 0$ such that $|f(x)| > A$ for all $x \in E$.

6.U. If $f \in L_p$, $1 \leqslant p \leqslant \infty$, and $g \in L_\infty$, then the product $fg \in L_p$ and $\|fg\|_p \leqslant \|f\|_p \|g\|_\infty$.

6.V. The space $L_\infty(X, \boldsymbol{X}, \mu)$ is contained in $L_1(X, \boldsymbol{X}, \mu)$ if and only if $\mu(X) < \infty$. If $\mu(X) = 1$ and $f \in L_\infty$, then

$$\|f\|_\infty = \lim_{p \to \infty} \|f\|_p.$$

CHAPTER 7

Modes of Convergence

We have already had occasion to mention four types of convergence of a sequence of measurable functions: pointwise convergence, almost everywhere convergence, uniform convergence, and convergence in L_p. There are two other notions of convergence that are of importance in dealing with measurable functions. We shall introduce these in this chapter and give interrelations between the various modes.

For convenience, we shall restate the definitions. **In this chapter we shall consider only real-valued functions defined on a fixed measure space** $(X, \mathbf{X}, \mu)$. In some applications it is necessary to consider extended real-valued functions, but this can usually be done by modifying the present discussion. In addition we shall limit our attention to L_p for $1 \leqslant p < \infty$, since the convergence L_∞ requires a special examination which is usually quite direct. Thus it will be understood that p is limited to these values.

The sequence (f_n) **converges uniformly** to f if for every $\varepsilon > 0$ there exists a natural number $N(\varepsilon)$ such that if $n \geqslant N(\varepsilon)$ and $x \in X$, then $|f_n(x) - f(x)| < \varepsilon$.

The sequence (f_n) **converges pointwise** to f if for every $\varepsilon > 0$ and $x \in X$ there is a natural number $N(\varepsilon, x)$, such that if $n \geqslant N(\varepsilon, x)$, then $|f_n(x) - f(x)| < \varepsilon$.

The sequence (f_n) **converges almost everywhere** to f if there exists a set M in $\mathbf{X}$ with $\mu(M) = 0$ such that for every $\varepsilon > 0$ and $x \in X \setminus M$

there exists a natural number $N(\varepsilon, x)$, such that if $n \geqslant N(\varepsilon, x)$, then $|f_n(x) - f(x)| < \varepsilon$.

It is obvious that uniform convergence implies pointwise convergence, that pointwise convergence implies almost everywhere convergence, and it is easily seen that the reverse implications do not hold. (Of course, if X consists of only a finite number of points, then pointwise convergence implies uniform convergence; if the only set with measure zero is the empty set, then almost everywhere convergence implies pointwise convergence.)

CONVERGENCE IN L_p

We now recall the notion of convergence in L_p, which was introduced in Chapter 6. We remark that an element in L_p is an equivalence class of functions which are real-valued and whose pth powers are integrable. However, by exercising some caution, we may regard an element of L_p as being a real-valued measurable function.

A sequence (f_n) in $L_p = L_p(X, \mathbf{X}, \mu)$ **converges in** L_p to $f \in L_p$, if for every $\varepsilon > 0$ there exists a natural number $N(\varepsilon)$ such that if $n \geqslant N(\varepsilon)$, then

$$\|f_n - f\|_p = \left\{\int |f_n - f|^p \, d\mu\right\}^{1/p} < \varepsilon.$$

In this case, we sometimes say that the sequence (f_n) **converges to** f **in mean (of order** p**)**.

A sequence (f_n) in L_p is said to be **Cauchy in** L_p, if for every $\varepsilon > 0$ there exists a natural number $N(\varepsilon)$ such that if $m, n \geqslant N(\varepsilon)$, then

$$\|f_m - f_n\|_p = \left\{\int |f_m - f_n|^p \, d\mu\right\}^{1/p} < \varepsilon.$$

We have seen in Theorem 6.14 that if (f_n) is Cauchy in L_p, then there exists an $f \in L_p$ such that (f_n) converges in L_p to f.

The relationship between convergence in L_p and the other modes of convergence that we have introduced is not so close. It is possible (see Exercise 7.A) for a sequence (f_n) in L_p to converge uniformly on X (and therefore pointwise and almost everywhere) to a function f in L_p, but not converge in L_p. However, if $\mu(X) < +\infty$, this cannot be the case.

7.1 THEOREM. *Suppose that $\mu(X) < +\infty$ and that (f_n) is a sequence in L_p which converges uniformly on X to f. Then f belongs to L_p and the sequence (f_n) converges in L_p to f.*

PROOF. Let $\varepsilon > 0$ and let $N(\varepsilon)$ be such that $|f_n(x) - f(x)| < \varepsilon$ whenever $n \geqslant N(\varepsilon)$ and $x \in X$. If $n \geqslant N(\varepsilon)$, then

$$\|f_n - f\|_p = \left\{ \int |f_n(x) - f(x)|^p \, d\mu \right\}^{1/p} \leqslant \left\{ \int \varepsilon^p \, d\mu \right\}^{1/p} = \varepsilon \mu(X)^{1/p}, \tag{7.1}$$

so that (f_n) converges in L_p to f. Q.E.D.

It is possible (see Exercise 7.B) for a sequence (f_n) in L_p to converge pointwise (and therefore almost everywhere) to a function f in L_p, but not converge in L_p even when $\mu(X) < +\infty$. However, if the sequence is dominated by a function in L_p, then the L_p convergence does take place.

7.2 THEOREM. *Let (f_n) be a sequence in L_p which converges almost everywhere to a measurable function f. If there exists a g in L_p such that*

$$|f_n(x)| \leqslant g(x), \qquad x \in X, \qquad n \in N, \tag{7.2}$$

then f belongs to L_p and (f_n) converges in L_p to f.

PROOF. In view of inequality (7.2), it follows that $|f(x)| \leqslant g(x)$ almost everywhere. Since $g \in L_p$, it follows from Corollary 5.4 that $f \in L_p$. Now

$$|f_n(x) - f(x)|^p \leqslant [2\, g(x)]^p, \text{ a.e.},$$

and since $\lim |f_n(x) - f(x)|^p = 0$, a.e., and $2^p g^p$ belongs to L_1, it follows from the Lebesgue Dominated Convergence Theorem 5.6 that

$$\lim \int |f_n - f|^p \, d\mu = 0.$$

Therefore (f_n) converges in L_p to f. Q.E.D.

7.3 COROLLARY. *Let $\mu(X) < +\infty$, and let (f_n) be a sequence in L_p which converges almost everywhere to a measurable function f. If there exists a constant K such that*

$$(7.3) \qquad |f_n(x)| \leqslant K, \quad x \in X, \quad n \in N,$$

then f belongs to L_p and (f_n) converges in L_p to f.

PROOF. If $\mu(X) < +\infty$, the constant functions belong to L_p. Q.E.D.

It might be suspected that convergence in L_p implies almost everywhere convergence, but this is not the case. In fact, we shall give an example of sequence (f_n) which converges in L_p to a function f, but such that $(f_n(x))$ does not converge to $f(x)$ for any x in X(!)

7.4 EXAMPLE. Let $X = [0, 1]$, $\boldsymbol{X} = \boldsymbol{B}$, and let λ be Lebesgue measure. We shall consider the intervals $[0, 1]$, $[0, \frac{1}{2}]$, $[\frac{1}{2}, 1]$, $[0, \frac{1}{3}]$, $[\frac{1}{3}, \frac{2}{3}]$, $[\frac{2}{3}, 1]$, $[0, \frac{1}{4}]$, $[\frac{1}{4}, \frac{1}{2}]$, $[\frac{1}{2}, \frac{3}{4}]$, $[\frac{3}{4}, 1]$, $[0, \frac{1}{5}]$, $[\frac{1}{5}, \frac{2}{5}]$, ...

Let f_n be the characteristic function of the nth interval on this list and let f be identically zero. If $n \geqslant m(m + 1)/2(= 1 + 2 + \cdots + m)$, then f_n is a characteristic function of an interval whose measure is at most $1/m$. Hence

$$\|f_n - f\|_p{}^p = \int |f_n - f|^p \, d\lambda$$
$$= \int f_n \, d\lambda \leqslant 1/m.$$

Therefore (f_n) converges in L_p to f. However, if x is any point of $[0, 1]$, then the sequence $(f_n(x))$ has a subsequence consisting only of 1's and another subsequence consisting only of 0's. Therefore, the sequence (f_n) *does not converge at any point of* $[0, 1]$. (It may be observed, however, that one can select a subsequence of (f_n) which converges to f.)

CONVERGENCE IN MEASURE

Although convergence in L_p does not imply almost everywhere convergence, it does imply another type of convergence that is often of interest.

7.5 DEFINITION. A sequence (f_n) of measurable real-valued functions is said to **converge in measure** to a measurable real-valued function f in case

$$\lim_{n\to\infty} \mu(\{x \in X : |f_n(x) - f(x)| \geqslant \alpha\}) = 0 \tag{7.4}$$

for each $\alpha > 0$. The sequence (f_n) is said to be **Cauchy in measure** in case

$$\lim_{m,n\to\infty} \mu(\{x \in X : |f_m(x) - f_n(x)| \geqslant \alpha\}) = 0 \tag{7.5}$$

for each $\alpha > 0$.

If (f_n) converges uniformly to f, then the set

$$\{x \in X : |f_n(x) - f(x)| \geqslant \alpha\}$$

is empty for sufficiently large n. Hence, uniform convergence implies convergence in measure. It is not difficult to show (see Exercise 7.D) that pointwise convergence (and therefore almost everywhere convergence) need not imply convergence in measure, unless the space X has finite measure (see Theorem 7.12). We observe, however, that convergence in L_p does imply convergence in measure. Indeed if $E_n(\alpha) = \{x \in X : |f_n(x) - f(x)| \geqslant \alpha\}$, then

$$\int |f_n - f|^p \, d\mu \geqslant \int_{E_n(\alpha)} |f_n - f|^p \, d\mu \geqslant \alpha^p \mu(E_n(\alpha)).$$

Since $\alpha > 0$, it follows that $\|f_n - f\|_p \to 0$ implies that $\mu(E_n(\alpha)) \to 0$ as $n \to \infty$.

The reader can readily verify that Example 7.4 also shows that a sequence can converge in measure to a function but not converge at any point. Despite that fact, we shall now prove a result due to F. Riesz that implies that if a sequence (f_n) converges in measure to f, then some subsequence converges almost everywhere to f. Actually we shall prove somewhat more than that.

7.6 THEOREM. *Let (f_n) be a sequence of measurable real-valued functions which is Cauchy in measure. Then there is a subsequence which converges almost everywhere and in measure to a measurable real-valued function f.*

PROOF. Select a subsequence (g_k) of (f_n) such that the set $E_k = \{x \in X : |g_{k+1}(x) - g_k(x)| \geqslant 2^{-k}\}$ is such that $\mu(E_k) < 2^{-k}$. Let $F_k = \bigcup_{j=k}^{\infty} E_j$ so that $F_k \in \boldsymbol{X}$ and $\mu(F_k) < 2^{-(k-1)}$. If $i \geqslant j \geqslant k$ and $x \notin F_k$, then

$$(7.6) \quad |g_i(x) - g_j(x)| \leqslant |g_i(x) - g_{i-1}(x)| + \cdots + |g_{j+1}(x) - g_j(x)| \leqslant \frac{1}{2^{i-1}} + \cdots + \frac{1}{2^j} < \frac{1}{2^{j-1}}.$$

Let $F = \bigcap_{k=1}^{\infty} F_k$ so that $F \in \boldsymbol{X}$ and $\mu(F) = 0$. From the argument just given it follows that (g_j) converges on $X \setminus F$. If we define f by

$$\begin{aligned} f(x) &= \lim g_j(x), & x &\notin F, \\ &= 0, & x &\in F, \end{aligned}$$

then (g_j) converges almost everywhere to the measurable real-valued function f. Passing to the limit in (7.6) as $i \to \infty$, we infer that if $j \geqslant k$ and $x \notin F_k$, then

$$|f(x) - g_j(x)| \leqslant \frac{1}{2^{j-1}} \leqslant \frac{1}{2^{k-1}}.$$

This shows that the sequence (g_j) converges uniformly to f on the complement of each set F_k.

To see that (g_j) converges in measure to f, let α, ε be positive real numbers and choose k so large that $\mu(F_k) < 2^{-(k-1)} < \inf(\alpha, \varepsilon)$. If $j \geqslant k$, the above estimate shows that

$$\begin{aligned} \{x \in X : |f(x) - g_j(x)| \geqslant \alpha\} &\subseteq \{x \in X : |f(x) - g_j(x)| > 2^{-(k-1)}\} \\ &\subseteq F_k. \end{aligned}$$

Therefore, $\mu(\{x \in X : |f(x) - g_j(x)| \geqslant \alpha\}) \leqslant \mu(F_k) < \varepsilon$ for all $j \geqslant k$, so that (g_j) converges in measure to f. Q.E.D.

7.7 COROLLARY. *Let (f_n) be a sequence of measurable real-valued functions which is Cauchy in measure. Then there is a measurable real-valued function f to which the sequence converges in measure. This limit function f is uniquely determined almost everywhere.*

PROOF. We have seen that there is a subsequence (f_{n_k}) which converges in measure to a function f. To see that the entire sequence converges in measure to f, observe that since

$$|f(x) - f_n(x)| \leqslant |f(x) - f_{n_k}(x)| + |f_{n_k}(x) - f_n(x)|,$$

it follows that

$$\{x \in X : |f(x) - f_n(x)| \geqslant \alpha\} \subseteq \left\{x \in X : |f(x) - f_{n_k}(x)| \geqslant \frac{\alpha}{2}\right\} \cup \left\{x \in X : |f_{n_k}(x) - f_n(x)| \geqslant \frac{\alpha}{2}\right\}.$$

The convergence in measure of (f_n) to f follows from this relation.

Suppose that the sequence (f_n) converges in measure to both f and g. Since

$$|f(x) - g(x)| \leqslant |f(x) - f_n(x)| + |f_n(x) - g(x)|,$$

it follows that

$$\{x \in X : |f(x) - g(x)| \geqslant \alpha\} \subseteq \left\{x \in X : |f(x) - f_n(x)| \geqslant \frac{\alpha}{2}\right\} \cup \left\{x \in X : |f_n(x) - g(x)| \geqslant \frac{\alpha}{2}\right\},$$

so that

$$\mu(\{x \in X : |f(x) - g(x)| \geqslant \alpha\}) = 0$$

for all $\alpha > 0$. Taking $\alpha = 1/n$, $n \in \boldsymbol{N}$, we infer that $f = g$, a.e. Q.E.D.

It has been remarked that convergence in L_p implies convergence in measure. In general, convergence in measure does not imply convergence in L_p (see Exercise 7.E). However, this implication does hold when the convergence is dominated.

7.8 THEOREM. *Let (f_n) be a sequence of functions in L_p which converges in measure to f and let $g \in L_p$ be such that*

$$|f_n(x)| \leqslant g(x), \qquad \text{a.e.}$$

Then $f \in L_p$ and (f_n) converges in L_p to f.

PROOF. If (f_n) does not converge in L_p to f, there exist a subsequence (g_k) of (f_n) and an $\varepsilon > 0$ such that

$$\|g_k - f\|_p > \varepsilon \qquad \text{for} \quad k \in \boldsymbol{N}. \tag{7.7}$$

Since (g_k) is a subsequence of (f_n), it follows (see Exercise 7.G) that it converges in measure to f. By Theorem 7.6 there is a subsequence (h_r) of (g_k) which converges almost everywhere and in measure to a function h. From the uniqueness part of Corollary 7.7 it follows that $h = f$ a.e. Since (h_r) converges almost everywhere to f and is dominated by g, Theorem 7.2 implies that $\|h_r - f\|_p \to 0$. However, this contradicts the relation (7.7). Q.E.D.

ALMOST UNIFORM CONVERGENCE

In the proof of Theorem 7.6 we constructed a sequence (g_j) of measurable real-valued functions which was uniformly convergent on the complement of sets which have arbitrarily small measure. At first mention this sounds equivalent to uniform convergence outside a set of zero measure, but it is not equivalent (see Exercise 7.J).

7.9 DEFINITION. A sequence (f_n) of measurable functions is said to be **almost uniformly convergent** to a measurable function f if for each $\delta > 0$ there is a set E_δ in $\boldsymbol{X}$ with $\mu(E_\delta) < \delta$ such that (f_n) converges uniformly to f on $X \setminus E_\delta$. The sequence (f_n) is said to be an **almost uniformly Cauchy sequence** if for every $\delta > 0$ there exists a set E_δ in $\boldsymbol{X}$ with $\mu(E_\delta) < \delta$ such that (f_n) is uniformly convergent on $X \setminus E_\delta$.

The reader is warned that the terminology (in addition to being unpleasant) is slightly at variance with the earlier use of the modifier "almost." It is clear that almost uniform convergence is implied by uniform convergence, but it is not hard to see that almost uniform convergence does not imply this stronger mode.

7.10 LEMMA. *Let (f_n) be an almost uniformly Cauchy sequence. Then there exists a measurable function f such that (f_n) converges almost uniformly and almost everywhere to f.*

PROOF. If $k \in N$, let $E_k \in \boldsymbol{X}$ be such that $\mu(E_k) < 2^{-k}$ and (f_n) is uniformly convergent on $X \setminus E_k$. Let $F_k = \bigcup_{j=k}^{\infty} E_j$, so that $F_k \in \boldsymbol{X}$ and $\mu(F_k) < 2^{-(k-1)}$. Note that (f_n) converges uniformly on $X \setminus F_k \subseteq X \setminus E_k$ and define g_k by

$$\begin{aligned} g_k(x) &= \lim f_n(x), && x \notin F_k, \\ &= 0, && x \in F_k. \end{aligned}$$

We observe that the sequence (F_k) is decreasing and that if $F = \bigcap F_k$, then $F \in \boldsymbol{X}$ and $\mu(F) = 0$. If $h \leqslant k$, then $g_h(x) = g_k(x)$ for all $x \in F_h$. Therefore, the sequence (g_k) converges on all of X to a measurable limit function which we shall denote by f. If $x \notin F_k$, then $f(x) = g_k(x) = \lim f_n(x)$. It follows that (f_n) converges to f on $X \setminus F$, so that (f_n) converges to f almost everywhere on X.

To see that the convergence is almost uniform, let $\varepsilon > 0$, and let K be so large that $2^{-(K-1)} < \varepsilon$. Then $\mu(F_K) < \varepsilon$, and (f_n) converges uniformly to $g_K = f$ on $X \setminus F_K$. Q.E.D.

The next result relates convergence in measure and almost uniform convergence.

7.11 THEOREM. *If a sequence (f_n) converges almost uniformly to f, then it converges in measure. Conversely, if a sequence (h_n) converges in measure to h, then some subsequence converges almost uniformly to h.*

PROOF. Suppose that (f_n) converges almost uniformly to f, and let α and ε be positive numbers. Then there exists a set E_ε in $\boldsymbol{X}$ with $\mu(E_\varepsilon) < \varepsilon$ such that (f_n) converges to f uniformly on $X \setminus E_\varepsilon$. Therefore, if n is sufficiently large, then the set $\{x \in X : |f_n(x) - f(x)| \geqslant \alpha\}$ must be contained in E_ε. This shows that (f_n) converges in measure to f.

Conversely, suppose that (h_n) converges in measure to h. It follows from Theorem 7.6 that there is a subsequence (g_k) of (h_n) which converges in measure to a function g and the proof of Theorem 7.6 actually shows that the convergence is almost uniform. Since (g_k) converges in measure to both h and g, it follows from Corollary 7.7 that $h = g$ a.e. Therefore the subsequence (g_k) of (h_n) converges almost uniformly to h.

Q.E.D.

It follows from the Theorem 7.11 that if a sequence converges in L_p, then it has a subsequence which converges almost uniformly. Conversely, it may be seen (see Exercise 7.K) that almost uniform convergence does not imply convergence in L_p in general, although it does if the convergence is dominated by a function in L_p (apply Theorem 7.8).

One of the consequences of Lemma 7.10 is that almost uniform convergence implies almost everywhere convergence. In general, the converse is false (see Exercise 7.L). However, it is a remarkable and important fact that if the functions are real-valued and if $\mu(X) < +\infty$, then almost everywhere convergence does imply almost uniform convergence.

7.12 EGOROFF'S THEOREM. *Suppose that $\mu(X) < +\infty$ and that (f_n) is a sequence of measurable real-valued functions which converges almost everywhere on X to a measurable real-valued function f. Then the sequence (f_n) converges almost uniformly and in measure to f.*

PROOF. We suppose without loss of generality that (f_n) converges at every point of X to f. If $m, n \in \boldsymbol{N}$, let

$$E_n(m) = \bigcup_{k=n}^{\infty} \left\{x \in X : |f_k(x) - f(x)| \geqslant \frac{1}{m}\right\},$$

so that $E_n(m)$ belongs to $\boldsymbol{X}$ and $E_{n+1}(m) \subseteq E_n(m)$. Since $f_n(x) \to f(x)$ for all $x \in X$, it follows that

$$\bigcap_{n=1}^{\infty} E_n(m) = \emptyset,$$

Since $\mu(X) < +\infty$, we infer that $\mu\big(E_n(m)\big) \to 0$ as $n \to +\infty$. If $\delta > 0$, let k_m be such that $\mu\big(E_{k_m}(m)\big) < \delta/2^m$ and let $E_\delta = \bigcup_{m=1}^{\infty} E_{k_m}(m)$, so that $E_\delta \in \boldsymbol{X}$ and $\mu(E_\delta) < \delta$. Observe that if $x \notin E_\delta$, then $x \notin E_{k_m}(m)$, so that

$$|f_k(x) - f(x)| < \frac{1}{m}$$

for all $k \geqslant k_m$. Therefore (f_k) is uniformly convergent on the complement of E_δ. Q.E.D.

It is convenient to have a table indicating the relations between the various modes of convergence we have been discussing. Modifying the idea in Reference [10], we present three diagrams relating almost everywhere convergence (denoted by AE), almost uniform convergence (denoted by AU), convergence in L_p (denoted by L_p), and convergence in measure (denoted by M). It is understood that in discussing L_p convergence, it is assumed that the functions belong to L_p. Diagram 7.1 pertains to the case of a general measure space. A

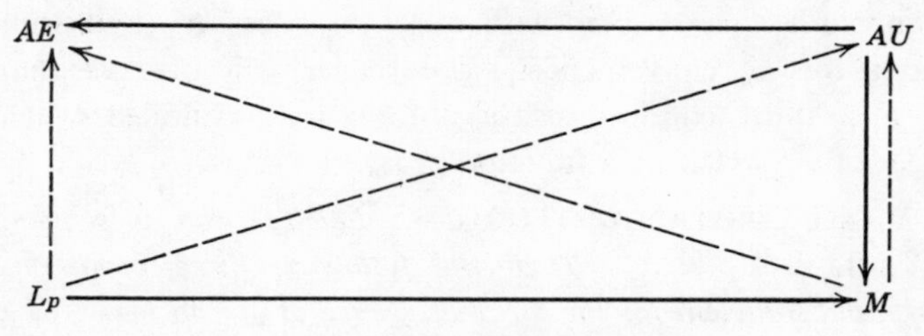

Diagram 7.1 General case

solid arrow signifies implication; a dashed arrow signifies that a subsequence converges in the indicated mode. The absence of an arrow indicates that a counterexample can be constructed. Diagram 7.2 relates to the case of a finite measure space. In view of Egoroff's

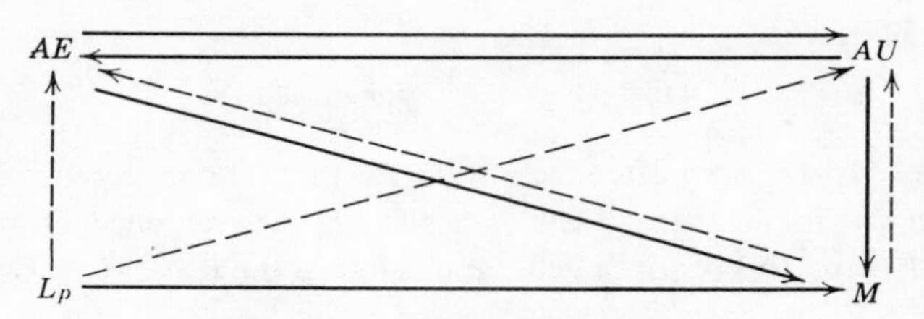

Diagram 7.2 Finite measure space

Theorem two implications are added. In Diagram 7.3, we assume that the sequence (f_n) is dominated by a function g in L_p. Here three implications are added.

We leave it as an exercise to verify that all the implications indicated in these diagrams hold, and that no other ones are valid without additional hypotheses.

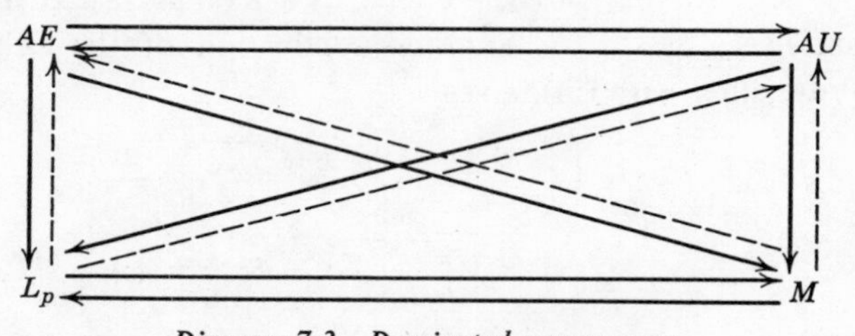

Diagram 7.3 Dominated convergence

We conclude this chapter with a set of necessary and sufficient conditions for L_p convergence. The reader will observe that the second and third conditions are automatically fulfilled when the sequence is dominated by a function in L_p.

7.13 VITALI CONVERGENCE THEOREM. *Let (f_n) be a sequence in $L_p(X, \boldsymbol{X}, \mu)$, $1 \leqslant p < \infty$. Then the following three conditions are necessary and sufficient for the L_p convergence of (f_n) to f:*

(i) *(f_n) converges to f in measure.*

(ii) *For each $\varepsilon > 0$ there is a set $E_\varepsilon \in \boldsymbol{X}$ with $\mu(E_\varepsilon) < +\infty$ such that if $F \in \boldsymbol{X}$ and $F \cap E_\varepsilon = \emptyset$, then*

$$\int_F |f_n|^p \, d\mu < \varepsilon^p \qquad \textit{for all } n \in \boldsymbol{N}.$$

(iii) *For each $\varepsilon > 0$ there is a $\delta(\varepsilon) > 0$, such that if $E \in \boldsymbol{X}$ and $\mu(E) < \delta(\varepsilon)$, then*

$$\int_E |f_n|^p \, d\mu < \varepsilon^p \qquad \textit{for all } n \in \boldsymbol{N}.$$

PROOF. It was seen after Definition 7.5 that L_p convergence implies convergence in measure. The fact that L_p convergence of the (f_n) implies (ii) and (iii) is not difficult and is left to the reader (see Exercises 6.R and 6.S).

We shall now show that these three conditions imply that (f_n) converges in L_p to f. If $\varepsilon > 0$, let E_ε be as in (ii) and let $F = X \setminus E_\varepsilon$ If the Minkowski Inequality is applied to $f_n - f_m = (f_n - f_m)\chi_{E_\varepsilon} + f_n \chi_F - f_m \chi_F$, we obtain

$$\|f_n - f_m\|_p \leqslant \left\{ \int_{E_\varepsilon} |f_n - f_m|^p \, d\mu \right\}^{1/p} + 2\varepsilon$$

for $n, m \in \boldsymbol{N}$. Now let $\alpha = \varepsilon[\mu(E_\varepsilon)]^{-1/p}$ and let $H_{nm} = \{x \in E_\varepsilon : |f_n(x) - f_m(x)| \geqslant \alpha\}$. In view of (i), there exists a $K(\varepsilon)$ such that if $n, m \geqslant K(\varepsilon)$, then $\mu(H_{nm}) < \delta(\varepsilon)$. Another application of the Minkowski Inequality together with (iii), gives

$$\begin{aligned} \left\{ \int_{E_\varepsilon} |f_n - f_m|^p \, d\mu \right\}^{1/p} &\leqslant \left\{ \int_{E_\varepsilon \setminus H_{nm}} |f_n - f_m|^p \, d\mu \right\}^{1/p} \\ &\quad + \left\{ \int_{H_{nm}} |f_n|^p \, d\mu \right\}^{1/p} + \left\{ \int_{H_{nm}} |f_m|^p \, d\mu \right\}^{1/p} \\ &\leqslant \alpha[\mu(E_\varepsilon)]^{1/p} + \varepsilon + \varepsilon = 3\varepsilon, \end{aligned}$$

when $n, m \geqslant K(\varepsilon)$. On combining this with the earlier inequality, we infer that the sequence (f_n) is Cauchy and hence convergent in L_p. Since we already know that (f_n) is convergent in measure to f, it follows from the uniqueness in Corollary 7.7 that (f_n) converges to f in L_p. Q.E.D.

EXERCISES

In these exercises $(\boldsymbol{R}, \boldsymbol{B}, \lambda)$ denotes the real line with Lebesgue measure defined on the Borel subsets of $\boldsymbol{R}$. Moreover, $1 \leqslant p < \infty$.

7.A. Let $f_n = n^{-1/p}\chi_{[0,n]}$. Show that the sequence (f_n) converges uniformly to the 0-function, but that it does not converge in $L_p(\boldsymbol{R}, \boldsymbol{B}, \lambda)$.

7.B. Let $f_n = n\,\chi_{[1/n,2/n]}$. Show that the sequence (f_n) converges everywhere to the 0-function but that it does not converge in $L_p(\boldsymbol{R}, \boldsymbol{B}, \lambda)$.

7.C. Show that both of the sequences in Exercises 7.A and 7.B converge in measure to their limits.

7.D. Let $f_n = \chi_{[n,n+1]}$. Show that the sequence (f_n) converges everywhere to the 0-function, but that it does not converge in measure.

7.E. The sequence in 7.B shows that convergence in measure does not imply L_p-convergence, even for a finite measure space.

7.F. Write down a subsequence of the sequence in Example 7.4 which converges almost everywhere to the 0-function. Can you find one which converges everywhere?

7.G. If a sequence (f_n) converges in measure to a function f, then every subsequence of (f_n) converges in measure to f. More generally, if (f_n) is Cauchy in measure, then every subsequence is Cauchy in measure.

7.H. If a sequence (f_n) converges in L_p to a function f, and a subsequence of (f_n) converges in L_p to g, then $f = g$ a.e.

7.I. If (f_n) is a sequence of characteristic functions of sets in $\boldsymbol{X}$, and if (f_n) converges to f in L_p, show that f is (almost everywhere equal to) the characteristic function of a set in $\boldsymbol{X}$.

7.J. Show that the sequence (f_n) in Exercise 7.B has the property that if $\delta > 0$, then it is uniformly convergent on the complement of the set $[0, \delta]$. However, show that there does not exist a set of measure zero, on the complement of which (f_n) is uniformly convergent.

7.K. Show that the sequence in Exercise 7.B converges almost uniformly but not in L_p.

7.L. Show that the sequence in Exercise 7.D converges everywhere, but not almost uniformly.

7.M. Let $f_n = n\,\chi_{[0,1]}$. Show that the hypothesis that the limit function be finite (at least almost everywhere) cannot be dropped in Egoroff's Theorem.

7.N. Show that Fatou's Lemma holds if almost everywhere convergence is replaced by convergence in measure.

7.O. Show that the Lebesgue Dominated Convergence Theorem holds if almost everywhere convergence is replaced by convergence in measure.

7.P. If $g \in L_p$ and $|f_n| \leqslant g$, show that conditions (ii) and (iii) of the Vitali Convergence Theorem 7.13 are satisfied.

7.Q. Let $(X, \boldsymbol{X}, \mu)$ be a finite measure space. If f is an $\boldsymbol{X}$-measurable function, let

$$r(f) = \int \frac{|f|}{1 + |f|}\, d\mu.$$

Show that a sequence (f_n) of $\boldsymbol{X}$-measurable functions converges in measure to f if and only if $r(f_n - f) \to 0$.

7.R. If the sequence (f_n) of measurable functions converges almost everywhere to a measurable function f and φ is continuous on $\boldsymbol{R}$ to $\boldsymbol{R}$, then the sequence $(\varphi \circ f_n)$ converges almost everywhere to $\varphi \circ f$. Conversely, if φ is not continuous at every point, then there exists a sequence (f_n) which converges almost everywhere to f but such that $(\varphi \circ f_n)$ does not converge almost everywhere to $\varphi \circ f$.

7.S. If φ is uniformly continuous on $\boldsymbol{R}$ to $\boldsymbol{R}$, and if (f_n) converges uniformly (respectively, almost uniformly, in measure) to f, then $(\varphi \circ f_n)$ converges uniformly (respectively, almost uniformly, in measure) to $\varphi \circ f$. Conversely, if φ is not uniformly continuous, there exists a measure space and a sequence (f_n) converging uniformly (and hence almost uniformly and in measure) to f but such that $(\varphi \circ f_n)$ does not converge in measure (and hence not uniformly or almost uniformly) to $\varphi \circ f$.

7.T. Let $(X, \boldsymbol{X}, \mu)$ be a finite measure space and let $1 \leqslant p < \infty$. Let φ be continuous on $\boldsymbol{R}$ to $\boldsymbol{R}$ and satisfy the condition: $(*)$ there exists $K > 0$ such that $|\varphi(t)| \leqslant K|t|$ for $|t| \geqslant K$. Show that $\varphi \circ f$ belongs to L_p for each $f \in L_p$. Conversely, if φ does not satisfy $(*)$, then there is a function f in L_p on a finite measure space such that $\varphi \circ f$ does not belong to L_p.

7.U. If (f_n) converges to f in L_p on a finite measure space, and if φ is continuous and satisfies condition $(*)$ of Exercise 7.T, then $(\varphi \circ f_n)$ converges in L_p to $\varphi \circ f$. Conversely, if condition $(*)$ is not satisfied, there exists a finite measure space and a sequence (f_n) which converges in L_p to f but such that $(\varphi \circ f_n)$ does not converge in L_p to $\varphi \circ f$.

7.V. Let $(X, \boldsymbol{X}, \mu)$ be an arbitrary measure space. Let φ be continuous on $\boldsymbol{R}$ to $\boldsymbol{R}$ and satisfy: $(**)$ there exists $K \geqslant 0$ such that $|\varphi(t)| \leqslant K|t|$ for all $t \in \boldsymbol{R}$. If $f \in L_p$, then $\varphi \circ f$ belongs to L_p. Conversely, if φ does not satisy $(**)$, there exists a measure space and a function $f \in L_p$ such that $\varphi \circ f$ does not belong to L_p.

7.W. If (f_n) converges to f in L_p on an arbitrary measure space, and if φ is continuous and satisfies $(**)$, then $(\varphi \circ f_n)$ converges to $\varphi \circ f$ in L_p. Conversely, if φ does not satisfy $(**)$, there exists a measure space and a sequence (f_n) which converges in L_p to f, but such that $(\varphi \circ f_n)$ does not converge in L_p to $\varphi \circ f$.

CHAPTER 8

Decomposition of Measures

In this chapter we shall consider the possibility of decomposing measures and charges in various ways and shall obtain some very useful results. First we shall consider charges and show that a charge can be written as the difference of two finite measures.

We recall from Definition 3.6 that a charge on a measurable space $(X, \mathbf{X})$ is a real-valued function λ defined on the σ-algebra $\mathbf{X}$ such that $\lambda(\emptyset) = 0$ and which is countably additive in the sense that

$$\lambda\left(\bigcup_{n=1}^{\infty} E_n\right) = \sum_{n=1}^{\infty} \lambda(E_n)$$

for any disjoint sequence (E_n) of sets in $\mathbf{X}$. The reader can easily check the proofs of Lemmas 3.3 and 3.4 to show that if (E_n) is an increasing sequence of sets in $\mathbf{X}$, then

$$\lambda\left(\bigcup_{n=1}^{\infty} E_n\right) = \lim \lambda(E_n), \tag{8.1}$$

and if (F_n) is a decreasing sequence of sets in $\mathbf{X}$, then

$$\lambda\left(\bigcap_{n=1}^{\infty} F_n\right) = \lim \lambda(F_n). \tag{8.2}$$

8.1 DEFINITION. If λ is a charge on $\mathbf{X}$, then a set P in $\mathbf{X}$ is said to be **positive** with respect to λ if $\lambda(E \cap P) \geqslant 0$ for any E in $\mathbf{X}$. A set N in $\mathbf{X}$ is said to be **negative** with respect to λ if $\lambda(E \cap N) \leqslant 0$ for any E in $\mathbf{X}$. A set M in $\mathbf{X}$ is said to be a **null set** for λ if $\lambda(E \cap M) = 0$ for any E in $\mathbf{X}$.

It is an exercise to show that a measurable subset of a positive set is positive and that the union of two positive sets is a positive set.

8.2 HAHN DECOMPOSITION THEOREM. *If λ is a charge on $\boldsymbol{X}$, then there exist sets P and N in $\boldsymbol{X}$ with $X = P \cup N$, $P \cap N = \emptyset$, and such that P is positive and N is negative with respect to λ.*

PROOF. The class $\boldsymbol{P}$ of all positive sets is not empty since it must contain $\emptyset$, at least. Let $\alpha = \sup\{\lambda(A) : A \in \boldsymbol{P}\}$, let (A_n) be a sequence in $\boldsymbol{P}$ such that $\lim \lambda(A_n) = \alpha$, and let $P = \bigcup_{n=1}^{\infty} A_n$. Since the union of two positive sets is positive, the sequence (A_n) can be chosen to be monotone increasing, and we shall assume that this has been done. Clearly P is a positive set for λ, since

$$\lambda(E \cap P) = \lambda\left(E \cap \bigcup_{n=1}^{\infty} A_n\right) = \lambda\left(\bigcup_{n=1}^{\infty} (E \cap A_n)\right) = \lim \lambda(E \cap A_n) \geqslant 0.$$

Moreover, $\alpha = \lim \lambda(A_n) = \lambda(P) < \infty$.

We shall now show that the set $N = X \setminus P$ is a negative set. If not, there is a measurable subset E of N with $\lambda(E) > 0$. The set E cannot be a positive set, for then $P \cup E$ would be a positive set with $\lambda(P \cup E) > \alpha$, contrary to the definition of α. Hence E contains sets with negative charge; let n_1 be the smallest natural number such that E contains a set E_1 in $\boldsymbol{X}$, such that $\lambda(E_1) \leqslant -1/n_1$. Now

$$\lambda(E \setminus E_1) = \lambda(E) - \lambda(E_1) > \lambda(E) > 0;$$

however, $E \setminus E_1$ cannot be a positive set, for then $P_1 = P \cup (E \setminus E_1)$ would be a positive set with $\lambda(P_1) > \alpha$. Therefore $E \setminus E_1$ contains sets with negative charge. Let n_2 be the smallest natural number such that $E \setminus E_1$ contains a set E_2 in $\boldsymbol{X}$ such that $\lambda(E_2) \leqslant -1/n_2$. As before $E \setminus (E_1 \cup E_2)$ is not a positive set, and we let n_3 be the smallest natural number such that $E \setminus (E_1 \cup E_2)$ contains a set E_3 in $\boldsymbol{X}$ such that $\lambda(E_3) \leqslant -1/n_3$. Repeating this argument, we obtain a disjoint sequence (E_k) of sets of $\boldsymbol{X}$ such that $\lambda(E_k) \leqslant -1/n_k$. Let $F = \bigcup_{k=1}^{\infty} E_k$ so that

$$\lambda(F) = \sum_{k=1}^{\infty} \lambda(E_k) \leqslant -\sum_{k=1}^{\infty} \frac{1}{n_k} \leqslant 0,$$

which shows that $1/n_k \to 0$. If G is a measurable subset of $E \setminus F$ and

$\lambda(G) < 0$, then $\lambda(G) < -1/(n_k - 1)$ for sufficiently large k, contradicting the fact that n_k is the smallest natural number such that $E \setminus (E_1 \cup \cdots \cup E_k)$ contains a set with charge less than $-1/n_k$. Hence, every measurable subset G of $E \setminus F$ must have $\lambda(G) \geqslant 0$, so that $E \setminus F$ is a positive set for λ. Since $\lambda(E \setminus F) = \lambda(E) - \lambda(F) > 0$, we infer that $P \cup (E \setminus F)$ is a positive set with charge exceeding α, which is a contradiction.

Therefore, it follows that the set $N = X \setminus P$ is a negative set for λ, and the desired decomposition of X is obtained. Q.E.D.

A pair P, N of measurable sets satisfying the conclusions of the preceding theorem is said to form a **Hahn decomposition** of X with respect to λ. In general, there will be no unique Hahn decomposition. In fact, if P, N is a Hahn decomposition for λ, and if M is a null set for λ, then $P \cup M$, $N \setminus M$ and $P \setminus M$, $N \cup M$ are also Hahn decompositions for λ. This lack of uniqueness is not an important matter for most purposes, however.

8.3 Lemma. *If P_1, N_1 and P_2, N_2 are Hahn decompositions for λ, and E belongs to $\mathbf{X}$, then*

$$\lambda(E \cap P_1) = \lambda(E \cap P_2), \qquad \lambda(E \cap N_1) = \lambda(E \cap N_2).$$

PROOF. Since $E \cap (P_1 \setminus P_2)$ is contained in the positive set P_1 and in the negative set N_2, then $\lambda\big(E \cap (P_1 \setminus P_2)\big) = 0$ so that

$$\lambda(E \cap P_1) = \lambda(E \cap P_1 \cap P_2).$$

Similarly,

$$\lambda(E \cap P_2) = \lambda(E \cap P_1 \cap P_2),$$

from which it follows that

$$\lambda(E \cap P_1) = \lambda(E \cap P_2).$$

Q.E.D.

8.4 Definition. Let λ be a charge on $\mathbf{X}$ and let P, N be a Hahn decomposition for λ. The **positive** and the **negative variations** of λ are the finite measures λ^+, λ^- defined for E in $\mathbf{X}$ by

$$\lambda^+(E) = \lambda(E \cap P), \qquad \lambda^-(E) = -\lambda(E \cap N). \tag{8.3}$$

The **total variation** of λ is the measure $|\lambda|$ defined for E in $\mathbf{X}$ by

$$|\lambda|(E) = \lambda^+(E) - \lambda^-(E).$$

It is a consequence of Lemma 8.3 that the positive and negative variations are well-defined and do not depend on the Hahn decomposition. It is also clear that

$$\lambda(E) = \lambda(E \cap P) + \lambda(E \cap N) = \lambda^+(E) - \lambda^-(E). \tag{8.4}$$

We shall state this result formally.

8.5 Jordan Decomposition Theorem. *If λ is a charge on $\mathbf{X}$, it is the difference of two finite measures on $\mathbf{X}$. In particular, λ is the difference of λ^+ and λ^-. Moreover, if $\lambda = \mu - \nu$ where μ, ν are finite measures on $\mathbf{X}$, then*

$$\mu(E) \geqslant \lambda^+(E), \qquad \nu(E) \geqslant \lambda^-(E) \tag{8.5}$$

for all E in $\mathbf{X}$.

PROOF. The representation $\lambda = \lambda^+ - \lambda^-$ has already been established. Since μ and ν have nonnegative values, then

$$\lambda^+(E) = \lambda(E \cap P) = \mu(E \cap P) - \nu(E \cap P)$$
$$\leqslant \mu(E \cap P) \leqslant \mu(E).$$

Similarly, $\lambda^-(E) \leqslant \nu(E)$ for any E in $\mathbf{X}$. Q.E.D.

We have seen, in Lemma 5.2, that if a function f is integrable with respect to a measure μ on $\mathbf{X}$, and if λ is defined for E in $\mathbf{X}$ by

$$\lambda(E) = \int_E f \, d\mu, \tag{8.6}$$

then λ is a charge. We now identify the positive and negative variations of λ.

8.6 Theorem. *If f belongs to $L(X, \mathbf{X}, \mu)$, and λ is defined by equation (8.6), then λ^+, λ^-, and $|\lambda|$ are given for E in $\mathbf{X}$ by*

$$\lambda^+(E) = \int_E f^+ \, d\mu, \quad \lambda^-(E) = \int_E f^- \, d\mu,$$
$$|\lambda|(E) = \int_E |f| \, d\mu.$$

PROOF. Let $P_f = \{x \in X : f(x) \geqslant 0\}$ and $N_f = \{x \in X : f(x) < 0\}$. Then $X = P_f \cup N_f$ and $P_f \cap N_f = \emptyset$. If $E \in \mathbf{X}$, then it is clear that $\lambda(E \cap P_f) \geqslant 0$ and $\lambda(E \cap N_f) \leqslant 0$. Hence P_f, N_f is a Hahn decomposition for λ. The statement now follows. Q.E.D.

It was seen in Corollary 4.9 that if f is a nonnegative extended real-valued measurable function and μ is a measure on $\mathbf{X}$, then the function λ defined by equation (8.6) is a measure on $\mathbf{X}$. There is a very important converse to this which gives conditions under which one can express a measure λ as an integral with respect to μ of a nonnegative extended real-valued measurable function. It was seen in Corollary 4.11 that a necessary condition for this representation is that $\lambda(E) = 0$ for any set E in $\mathbf{X}$ for which $\mu(E) = 0$. It turns out this condition is also sufficient in the important case where λ and μ are σ-finite.

8.7 DEFINITION. A measure λ on $\mathbf{X}$ is said to be **absolutely continuous** with respect to a measure μ on $\mathbf{X}$ if $E \in \mathbf{X}$ and $\mu(E) = 0$ imply that $\lambda(E) = 0$. In this case we write $\lambda \ll \mu$. A charge λ is **absolutely continuous** with respect to a charge μ in case the total variation $|\lambda|$ of λ is absolutely continuous with respect to $|\mu|$.

The following lemma is useful and adds to our intuitive understanding of absolute continuity.

8.8 LEMMA. *Let λ and μ be finite measures on $\mathbf{X}$. Then $\lambda \ll \mu$ if and only if for every $\varepsilon > 0$ there exists a $\delta(\varepsilon) > 0$ such that $E \in \mathbf{X}$ and $\mu(E) < \delta(\varepsilon)$ imply that $\lambda(E) < \varepsilon$.*

PROOF. If this condition is satisfied and $\mu(E) = 0$, then $\lambda(E) < \varepsilon$ for all $\varepsilon > 0$, from which it follows that $\lambda(E) = 0$.

Conversely, suppose that there exist an $\varepsilon > 0$ and sets $E_n \in \mathbf{X}$ with $\mu(E_n) < 2^{-n}$ and $\lambda(E_n) \geqslant \varepsilon$. Let $F_n = \bigcup_{k=n}^{\infty} E_k$, so that $\mu(F_n) < 2^{-n+1}$ and $\lambda(F_n) \geqslant \varepsilon$. Since (F_n) is a decreasing sequence of measurable sets,

$$\mu\left(\bigcap_{n=1}^{\infty} F_n\right) = \lim \mu(F_n) = 0,$$

$$\lambda\left(\bigcap_{n=1}^{\infty} F_n\right) = \lim \lambda(F_n) \geqslant \varepsilon.$$

Hence λ is not absolutely continuous with respect to μ. Q.E.D.

8.9 RADON-NIKODÝM THEOREM. *Let λ and μ be σ-finite measures defined on $\boldsymbol{X}$ and suppose that λ is absolutely continuous with respect to μ. Then there exists a function f in $M^+(X, \boldsymbol{X})$ such that*

$$\lambda(E) = \int_E f\,d\mu, \qquad E \in \boldsymbol{X}. \tag{8.6}$$

Moreover, the function f is uniquely determined μ-almost everywhere.

PROOF. We shall first prove Theorem 8.9 under the hypothesis that $\lambda(X)$ and $\mu(X)$ are finite.

If $c > 0$, let $P(c)$, $N(c)$ be a Hahn decomposition of X for the charge $\lambda - c\mu$. If $k \in \boldsymbol{N}$, consider the measurable sets

$$A_1 = N(c), \qquad A_{k+1} = N\big((k+1)c\big) \setminus \bigcup_{j=1}^{k} A_j.$$

It is clear that the sets A_k, $k \in \boldsymbol{N}$, are disjoint and that

$$\bigcup_{j=1}^{k} N(jc) = \bigcup_{j=1}^{k} A_j.$$

It follows that

$$A_k = N(kc) \setminus \bigcup_{j=1}^{k-1} N(jc) = N(kc) \cap \bigcap_{j=1}^{k-1} P(jc).$$

Hence if E is a measurable subset of A_k, then $E \subseteq N(kc)$ and $E \subseteq P\big((k-1)c\big)$ so that

$$(k-1)c\mu(E) \leqslant \lambda(E) \leqslant kc\mu(E). \tag{8.7}$$

Define B by

$$B = X \setminus \bigcup_{j=1}^{\infty} A_j = \bigcap_{j=1}^{\infty} P(jc),$$

so that $B \subseteq P(kc)$ for all $k \in \boldsymbol{N}$. This implies that

$$0 \leqslant kc\mu(B) \leqslant \lambda(B) \leqslant \lambda(X) < +\infty$$

for all $k \in \boldsymbol{N}$, so that $\mu(B) = 0$. Since $\lambda \ll \mu$, we infer that $\lambda(B) = 0$.

Let f_c be defined by $f_c(x) = (k-1)c$ for $x \in A_k$ and $f_c(x) = 0$ for $x \in B$. If E is an arbitrary measurable set, then E is the union of the disjoint sets $E \cap B$, $E \cap A_k$, $k \in \boldsymbol{N}$, so it follows from (8.7) that

$$\int_E f_c\,d\mu \leqslant \lambda(E) \leqslant \int_E (f_c + c)\,d\mu \leqslant \int_E f_c\,d\mu + c\mu(X).$$

We employ the preceding construction for $c = 2^{-n}$, $n \in N$, to obtain a sequence of functions we now denote by f_n. Hence

$$\int_E f_n \, d\mu \leqslant \lambda(E) \leqslant \int_E f_n \, d\mu + 2^{-n}\mu(X). \tag{8.8}$$

for all $n \in N$. Let $m \geqslant n$, and observe that

$$\int_E f_n \, d\mu \leqslant \lambda(E) \leqslant \int_E f_m \, d\mu + 2^{-m} \mu(X),$$

$$\int_E f_m \, d\mu \leqslant \lambda(E) \leqslant \int_E f_n \, d\mu + 2^{-n} \mu(X),$$

from which it is seen that

$$\left| \int_E (f_n - f_m) \, d\mu \right| \leqslant 2^{-n} \mu(X),$$

for all E in X. If we let E be the sets where the integrand is positive and negative and combine, we deduce that

$$\int |f_n - f_m| \, d\mu \leqslant 2^{-n+1} \mu(X)$$

whenever $m \geqslant n$. Thus the sequence (f_n) converges in mean to a function f. Since the f_n belong to M^+, it is clear from Theorem 7.6 that we may require that $f \in M^+$. Moreover,

$$\left| \int_E f_n \, d\mu - \int_E f \, d\mu \right| \leqslant \int_E |f_n - f| \, d\mu \leqslant \int |f_n - f| \, d\mu,$$

so that we conclude from (8.8) that

$$\lambda(E) = \lim \int_E f_n \, d\mu = \int_E f \, d\mu$$

for all $E \in X$. This completes the proof of the existence assertion of the theorem in the case where both λ and μ are finite measures.

We claim that f is uniquely determined up to sets of μ-measure zero. Indeed, suppose that $f, h \in M^+$ and that

$$\lambda(E) = \int_E f \, d\mu = \int_E h \, d\mu$$

for all E in $\boldsymbol{X}$. Let $E_1 = \{x : f(x) > h(x)\}$ and $E_2 = \{x : f(x) < h(x)\}$, and apply Corollary 4.10 to infer that $f(x) = h(x)$ μ-almost everywhere.

We shall now suppose that λ and μ are σ-finite and let (X_n) be an increasing sequence of sets in $\boldsymbol{X}$ such that

$$\lambda(X_n) < \infty, \qquad \mu(X_n) < \infty.$$

Apply the preceding argument to obtain a function h_n in M^+ which vanishes for $x \notin X_n$, such that if E is a measurable subset of X_n, then

$$\lambda(E) = \int_E h_n \, d\mu.$$

If $n \leqslant m$, then $X_n \subseteq X_m$, and it follows that

$$\int_E h_n \, d\mu = \int_E h_m \, d\mu$$

for any measurable subset E of X_n. From the uniqueness of h_n, it follows that $h_m(x) = h_n(x)$ for μ-almost all x in X_n whenever $m \geqslant n$. Let $f_n = \sup \{h_1, \ldots, h_n\}$ so that (f_n) is a monotone increasing sequence in M^+ and let $f = \lim f_n$. If $E \in \boldsymbol{X}$, then

$$\lambda(E \cap X_n) = \int_E f_n \, d\mu.$$

Since $(E \cap X_n)$ is an increasing sequence of sets with union E, it follows from Lemma 3.3 and the Monotone Convergence Theorem 4.6 that

$$\begin{aligned} \lambda(E) &= \lim \lambda(E \cap X_n) = \lim \int_E f_n \, d\mu \\ &= \int_E f \, d\mu. \end{aligned}$$

The μ-uniqueness of f is established as before. Q.E.D.

The function f whose existence we have established is often called the **Radon-Nikodým derivative** of λ with respect to μ, and is denoted by $d\lambda/d\mu$. It will be seen in the exercises to have properties closely related to the derivative. The reader should observe that this function is not necessarily integrable; in fact, f is (μ-equivalent to) an integrable function if and only if λ is a finite measure.

In intuitive terms, a measure λ is absolutely continuous with respect to a measure μ in case sets which have small μ-measure also have small λ-measure. At the opposite extreme, there is the notion of singular measures, which we now introduce.

8.10 DEFINITION. Two measures λ, μ on $\boldsymbol{X}$ are said to be **mutually singular** if there are disjoint sets A, B in $\boldsymbol{X}$ such that $X = A \cup B$ and $\lambda(A) = \mu(B) = 0$. In this case we write $\lambda \perp \mu$.

Although the relation of singularity is symmetric in λ and μ, we shall sometimes say that λ is **singular with respect to** μ.

8.11 LEBESGUE DECOMPOSITION THEOREM. *Let λ and μ be σ-finite measures defined on a σ-algebra $\boldsymbol{X}$. Then there exists a measure λ_1 which is singular with respect to μ and a measure λ_2 which is absolutely continuous with respect to μ such that $\lambda = \lambda_1 + \lambda_2$. Moreover, the measures λ_1 and λ_2 are unique.*

PROOF. Let $\nu = \lambda + \mu$ so that ν is a σ-finite measure. Since λ and μ are both absolutely continuous with respect to ν, the Radon-Nikodým Theorem implies that there exist functions f, g in $M^+(X, \boldsymbol{X})$ such that

$$\lambda(E) = \int_E f\,d\nu, \qquad \mu(E) = \int_E g\,d\nu$$

for all E in $\boldsymbol{X}$. Let $A = \{x : g(x) = 0\}$, and let $B = \{x : g(x) > 0\}$, so that $A \cap B = \emptyset$, and $X = A \cup B$.

Define λ_1 and λ_2 for E in $\boldsymbol{X}$ by

$$\lambda_1(E) = \lambda(E \cap A), \qquad \lambda_2(E) = \lambda(E \cap B).$$

Since $\mu(A) = 0$, it follows that $\lambda_1 \perp \mu$. To see that $\lambda_2 \ll \mu$, observe that if $\mu(E) = 0$, then

$$\int_E g\,d\nu = 0,$$

so that $g(x) = 0$ for ν-almost all x in E. Hence $\nu(E \cap B) = 0$; since $\lambda \ll \nu$,

$$\lambda_2(E) = \lambda(E \cap B) = 0.$$

Clearly $\lambda = \lambda_1 + \lambda_2$, so the existence of this decomposition is affirmed.

To establish the uniqueness of the decomposition, use the observation that if α is a measure such that $\alpha \ll \mu$, and $\alpha \perp \mu$, then $\alpha = 0$. Q.E.D.

RIESZ REPRESENTATION THEOREM

As another application of the Radon–Nikodým Theorem, we shall present theorems concerning the representation of bounded linear functionals on the spaces L_p, $1 \leqslant p < \infty$.

8.12 DEFINITION. A **linear functional** on $L_p = L_p(X, \boldsymbol{X}, \mu)$ is a mapping G of L_p into $\boldsymbol{R}$ such that

$$G(af + bg) = aG(f) + bG(g)$$

for all a, b in $\boldsymbol{R}$ and f, g in L_p. The linear functional G is **bounded** if there exists a constant M such that

$$|G(f)| \leqslant M\|f\|_p$$

for all f in L_p. In this case, the **bound** or the **norm** of G is defined to be

$$\|G\| = \sup\{|G(f)| : f \in L_p, \|f\|_p \leqslant 1\}. \tag{8.9}$$

It is a consequence of the linearity of the integral and Hölder's Inequality that if $g \in L_q$ (where $q = \infty$ when $p = 1$ and $q = p/(p-1)$ otherwise) and if we define G on L_p by

$$G(f) = \int fg \, d\mu, \tag{8.10}$$

then G is a linear functional with norm at most equal to $\|g\|_q$ (and it is an exercise to prove that $\|G\| = \|g\|_q$). The Riesz Theorem yields a converse to this observation.

Before we prove this theorem it is convenient to observe that any bounded linear functional on L_p can be written as the difference of two positive linear functionals (that is, functionals G such that $G(f) \geqslant 0$ for all $f \in L_p$ for which $f \geqslant 0$).

8.13 LEMMA. *Let G be a bounded linear functional on L_p. Then there exist two positive bounded linear functionals G^+, G^- such that $G(f) = G^+(f) - G^-(f)$ for all $f \in L_p$.*

PROOF. If $f \geqslant 0$ define $G^+(f) = \sup\{G(g) : g \in L_p,\ 0 \leqslant g \leqslant f\}$. It is clear that $G^+(cf) = c\,G^+(f)$ for $c \geqslant 0$ and $f \geqslant 0$. If $0 \leqslant g_j \leqslant f_j$, then

$$G(g_1) + G(g_2) = G(g_1 + g_2) \leqslant G^+(f_1 + f_2).$$

Taking the suprema over all such g_j in L_p we obtain $G^+(f_1) + G^+(f_2) \leqslant G^+(f_1 + f_2)$. Conversely, if $0 \leqslant h \leqslant f_1 + f_2$, let $g_1 = \sup(h - f_2, 0)$ and $g_2 = \inf(h, f_2)$. It follows that $g_1 + g_2 = h$ and that $0 \leqslant g_j \leqslant f_j$. Therefore $G(h) = G(g_1) + G(g_2) \leqslant G^+(f_1) + G^+(f_2)$; since this holds for all such $h \in L_p$, we infer that

$$G^+(f_1 + f_2) = G^+(f_1) + G^+(f_2)$$

for all f_j in L_p such that $f_j \geqslant 0$.

If f is an arbitrary element of L_p, define

$$G^+(f) = G^+(f^+) - G^+(f^-).$$

It is an elementary exercise to show that G^+ is a bounded linear functional on L_p. Further, we define G^- for $f \in L_p$ by

$$G^-(f) = G^+(f) - G(f),$$

so that G^- is evidently a bounded linear functional. From the definition of G^+ it is readily seen that G^- is a positive linear functional, and it is obvious that $G = G^+ - G^-$. Q.E.D.

8.14 Riesz Representation Theorem. *If $(X, \boldsymbol{X}, \mu)$ is a σ-finite measure space and G is a bounded linear functional on $L_1(X, \boldsymbol{X}, \mu)$, then there exists a g in L_∞ $(X, \boldsymbol{X}, \mu)$ such that equation* (8.10) *holds for all f in L_1. Moreover, $\|G\| = \|g\|_\infty$ and $g \geqslant 0$ if G is a positive linear functional.*

PROOF. We shall first suppose that $\mu(X) < \infty$ and that G is positive. Define λ on $\boldsymbol{X}$ to $\boldsymbol{R}$ by $\lambda(E) = G(\chi_E)$; clearly $\lambda(\emptyset) = 0$. If (E_n) is an increasing sequence in $\boldsymbol{X}$ and $E = \bigcup E_n$, then (χ_{E_n}) converges pointwise to χ_E. Since $\mu(X) < \infty$, it follows from Corollary 7.3 that this sequence converges in L_1 to χ_E. Since

$$\begin{aligned} 0 \leqslant \lambda(E) - \lambda(E_n) &= G(\chi_E) - G(\chi_{E_n}) \\ &= G(\chi_E - \chi_{E_n}) \leqslant \|G\| \, \|\chi_E - \chi_{E_n}\|_1, \end{aligned}$$

it follows that λ is a measure. Moreover, if $M \in \boldsymbol{X}$ and $\mu(M) = 0$, then $\lambda(M) = 0$, so that $\lambda \ll \mu$.

On applying the Radon-Nikodým Theorem we obtain a nonnegative measurable function on X to $\boldsymbol{R}$ such that

$$G(\chi_E) = \lambda(E) = \int \chi_E \, g \, d\mu$$

for all $E \in \boldsymbol{X}$. It follows by linearity that

$$G(\varphi) = \int \varphi\, g\, d\mu$$

for all $\boldsymbol{X}$-measurable simple functions φ.

If f is a nonnegative function in L_1, let (φ_n) be a monotone increasing sequence of simple functions converging almost everywhere and in L_1 to f. From the boundedness of G it is seen that $G(f) = \lim G(\varphi_n)$. Moreover, it follows from the Monotone Convergence Theorem that

$$G(f) = \lim_n \int \varphi_n\, g\, d\mu = \int fg\, d\mu .$$

This relation holds for arbitrary $f \in L_1$ by linearity.

We now turn to the σ-finite case. If $X = \bigcup F_n$, where (F_n) is an increasing sequence of sets in $\boldsymbol{X}$ with finite measure, the preceding argument yields the existence of nonnegative functions g_n such that

$$G(f\chi_{F_n}) = \int f\chi_{F_n} g_n\, d\mu$$

for all f in L_p. If $m \leqslant n$ it is readily seen that $g_m(x) = g_n(x)$ for almost all x in F_m. In this way we obtain a function g which represents G.

If G is an arbitrary bounded linear functional on L_1, Lemma 8.13 shows that we can write $G = G^+ - G^-$, where G^+ and G^- are bounded positive linear functionals. If we apply the preceding considerations to G^+ and G^-, we obtain nonnegative measurable functions g^+, g^- which represent G^+, G^-. If we set $g = g^+ - g^-$, we obtain the representation

$$G(f) = \int f g\, d\mu \tag{8.10}$$

for all $f \in L_1$. It will be left as an exercise to show that $\|G\| = \|g\|_\infty$.

Q.E.D.

8.15 Riesz Representation Theorem. *If $(X, \boldsymbol{X}, \mu)$ is an arbitrary measure space and G is a bounded linear functional on $L_p(X, \boldsymbol{X}, \mu)$, $1 < p < \infty$, then there exists a g in $L_q(X, \boldsymbol{X}, \mu)$, where $q = p/(p-1)$, such that equation* (8.10) *holds for all f in L_p. Moreover, $\|G\| = \|g\|_q$.*

PROOF. If $\mu(X) < \infty$, the proof of the preceding theorem requires only minor changes to show that there exists a g in L_q with $\|G\| = \|g\|_q$ and such that

$$G(f) = \int f g \, d\mu$$

for all f in L_p. In addition, the procedure used before applies to extend the result to the case where $(X, \boldsymbol{X}, \mu)$ is σ-finite.

We now complete the proof by observing that a bounded linear functional "vanishes off of a σ-finite set." More precisely, let (f_n) be a sequence in L_p such that $\|f_n\| = 1$ and

$$G(f_n) \geqslant \|G\| \left(1 - \frac{1}{n}\right).$$

There exists a σ-finite set X_0 in $\boldsymbol{X}$ outside of which all the f_n vanish. Let $E \in \boldsymbol{X}$ with $E \cap X_0 = \emptyset$, then $\|f_n \pm t\,\chi_E\|_p = (1 + t^p\,\mu(E))^{1/p}$ for $t \geqslant 0$. Moreover, since

$$G(f_n) - G(\pm\, t\,\chi_E) \leqslant |G(f_n \pm t\,\chi_E)|,$$

it follows that

$$|G(t\chi_E)| \leqslant \|G\| \left\{ \left(1 + t^p\,\mu(E)\right)^{1/p} - \left(1 - \frac{1}{n}\right) \right\}.$$

for all n in $\boldsymbol{N}$. First let $n \to \infty$, and then divide by $t > 0$, to get

$$|G(\chi_E)| \leqslant \|G\| \frac{(1 + t^p\,\mu(E))^{1/p} - 1}{t}.$$

If we apply L'Hospital's Rule as $t \to 0+$, we infer that $G(\chi_E) = 0$, for any $E \in \boldsymbol{X}$ in the complement of the σ-finite set X_0. Therefore, if f is any function in L_p such that $X_0 \cap \{x \in X : f(x) \neq 0\} = \emptyset$, it follows that $G(f) = 0$.

Hence we can apply the preceding argument to obtain a function g on X_0 which represents G, and extend g to all of X by requiring that it vanish on the complement of X_0. In this way we obtain the desired function. Q.E.D.

EXERCISES

8.A. If P is a positive set with respect to a charge λ, and if $E \in \boldsymbol{X}$ and $E \subseteq P$, then E is positive with respect to λ.

8.B. If P_1 and P_2 are positive sets for a charge λ, then $P_1 \cup P_2$ is positive for λ.

8.C. A set M in $\boldsymbol{X}$ is a null set for a charge λ if and only if $|\lambda|(M) = 0$.

8.D. If λ is a charge on $\boldsymbol{X}$, then the values of λ are bounded and

$$\lambda^+(E) = \sup\{\lambda(F) : F \subseteq E, F \in \boldsymbol{X}\},$$
$$\lambda^-(E) = -\inf\{\lambda(F) : F \subseteq E, F \in \boldsymbol{X}\}.$$

8.E. Let μ_1, μ_2, and μ_3 be measures on $(X, \boldsymbol{X})$. Show that $\mu_1 \ll \mu_1$ and that $\mu_1 \ll \mu_2$ and $\mu_2 \ll \mu_3$ imply that $\mu_1 \ll \mu_3$. Give an example to show that $\mu_1 \ll \mu_2$ does not imply that $\mu_2 \ll \mu_1$.

8.F. If (μ_n) is a sequence of measures on $(X, \boldsymbol{X})$ with $\mu_n(X) \leqslant 1$, let λ be defined for E in $\boldsymbol{X}$ by

$$\lambda(E) = \sum_{n=1}^{\infty} 2^{-n} \mu_n(E).$$

Show that λ is a measure and that $\mu_n \ll \lambda$ for all n.

8.G. Let λ be a charge and let μ be a measure on $(X, \boldsymbol{X})$. If $\lambda \ll \mu$, then λ^+, λ^-, and $|\lambda|$ are absolutely continuous with respect to μ.

8.H. Show that Lemma 8.8 is true even if μ is allowed to be an infinite measure. However, it may fail if λ is an infinite measure. [*Hint:* Let λ be the counting measure on $\boldsymbol{N}$, and let

$$\mu(E) = \sum_{n \in E} 2^{-n}.]$$

8.I. Let μ be defined as in Exercise 8.H and if $E \subseteq \boldsymbol{N}$, let λ be defined by

$$\begin{aligned} \lambda(E) &= 0, &&\text{if } E = \emptyset; \\ &= +\infty, &&\text{if } E \neq \emptyset. \end{aligned}$$

Show that μ is a finite measure on the σ-algebra $\boldsymbol{X}$ of all subsets of $\boldsymbol{N}$, and that λ is an infinite measure on $\boldsymbol{X}$. Moreover, $\lambda \ll \mu$ and $\mu \ll \lambda$.

8.J. If λ and μ are σ-infinite and $\lambda \ll \mu$, then the function f in the Radon-Nikodým Theorem can be taken to be finite-valued on X.

8.K. Let μ be a finite measure, let $\lambda \ll \mu$, and let P_n, N_n be a Hahn decomposition for $\lambda - n\mu$. Let $P = \bigcap P_n$, $N = \bigcup N_n$. Show that N is σ-finite for λ and that if $E \subseteq P$, $E \in \boldsymbol{X}$, then either $\lambda(E) = 0$ or $\lambda(E) = +\infty$.

8.L. Use Exercise 8.K to extend the Radon-Nikodým Theorem to the case where μ is σ-finite and λ is an arbitrary measure with $\lambda \ll \mu$. Here f is not necessarily finite-valued.

8.M. (a) Let X be an uncountable set and $\boldsymbol{X}$ be the family of all subsets E of X such that either E or $X \setminus E$ is countable. Let $\mu(E)$ equal the number of elements in E if E is finite and equal $+\infty$ otherwise, and let $\lambda(E) = 0$ if E is countable and equal $+\infty$ if E is uncountable. Then $\lambda \ll \mu$, but the Radon-Nikodým Theorem fails.

(b) Let $X = [0, 1]$ and let $\boldsymbol{X}$ be the Borel subsets of X. If μ is the counting measure on $\boldsymbol{X}$ and λ is Lebesgue measure on $\boldsymbol{X}$, then λ is a finite measure and $\lambda \ll \mu$, but the Radon-Nikodým Theorem fails.

8.N. Let λ, μ be σ-finite measures on $(X, \boldsymbol{X})$, let $\lambda \ll \mu$, and let $f = d\lambda/d\mu$. If g belongs to $M^+(X, \boldsymbol{X})$, then

$$\int g \, d\lambda = \int gf \, d\mu.$$

(*Hint:* First consider simple functions and apply the Monotone Convergence Theorem.)

8.O. Let λ, μ, ν be σ-finite measures on $(X, \boldsymbol{X})$. Use Exercise 8.N to show that if $\nu \ll \lambda$ and $\lambda \ll \mu$, then

$$\frac{d\nu}{d\mu} = \frac{d\nu}{d\lambda}\frac{d\lambda}{d\mu}, \quad \mu\text{-almost everywhere.}$$

Also, if $\lambda_j \ll \mu$ for $j = 1, 2$, then

$$\frac{d}{d\mu}(\lambda_1 + \lambda_2) = \frac{d\lambda_1}{d\mu} + \frac{d\lambda_2}{d\mu}, \quad \mu\text{-almost everywhere.}$$

8.P. If λ and μ are σ-finite, $\lambda \ll \mu$, and $\mu \ll \lambda$, then

$$\frac{d\lambda}{d\mu} = \frac{1}{d\mu/d\lambda}, \quad \text{almost everywhere.}$$

8.Q. If λ and μ are measures, with $\lambda \ll \mu$ and $\lambda \perp \mu$, then $\lambda = 0$.

8.R. If λ is a charge and μ is a measure, then $\lambda \perp \mu$ implies that λ^+, λ^-, $|\lambda|$ are singular with respect to μ.

8.S. The collection of all charges on $(X, \mathbf{X})$ is a Banach space under the vector operations

$$(c\mu)(E) = c\mu(E), \quad (\lambda + \mu)(E) = \lambda(E) + \mu(E)$$

and the norm $\|\mu\| = |\mu|(X)$.

8.T. Suppose g satisfies equation (8.10) for all f in L_1 and that $c > 1$. Let $E_c = \{x : |g(x)| \geqslant c\|G\|\}$, and define $f_c(x)$ to be ± 1 when $\pm g(x) \geqslant c\|G\|$ and to be 0 when $x \notin E_c$. Then

$$c\|G\|\mu(E_c) \leqslant G(f_c) \leqslant \|G\|\mu(E_c),$$

which is a contradiction unless $\mu(E_c) = 0$. Infer that $|g(x)| \leqslant \|G\|$ for μ-almost all x.

8.U. If g satisfies (8.10) for all $f \in L_p$, show that $g \in L_q$ and that $\|G\| = \|g\|_q$.

8.V. The Riesz Representation Theorem for $p = 2$ can be proved by some elementary Hilbert space geometry (see [5], pp. 249–50). We now show that this result can be used to prove the Radon-Nikodým Theorem. We shall limit our attention to finite measures λ, μ with $\lambda \ll \mu$. Let $\nu = \lambda + \mu$ and show that

$$G(f) = \int f \, d\lambda$$

defines a positive linear functional on $L_2(X, \mathbf{X}, \nu)$ with norm at most 1. If $g \in L_2(X, \mathbf{X}, \nu)$ is such that

$$G(f) = \int f g \, d\nu, \qquad f \in L_2(X, \mathbf{X}, \nu),$$

then we see by taking $f = \chi_E$, $E \in \mathbf{X}$, that $0 \leqslant g(x) \leqslant 1$ for ν-almost all x. Moreover, $\mu\{x : g(x) = 1\} = 0$. Since $\nu = \lambda + \mu$, we have

$$\int h\,(1 - g)\, d\lambda = \int h\, g \, d\mu$$

for all nonnegative $h \in L_2(X, \mathbf{X}, \nu)$ and hence for all nonnegative measurable h. Now take $h = \chi_E/(1 - g)$ to infer that

$$\lambda(E) = \int_E \left(\frac{g}{1 - g}\right) d\mu.$$

CHAPTER 9

Generation of Measures

In the preceding chapters we have given a few examples of measures, but they are of a rather special form, and it is time to demonstrate how measures can be constructed. In particular, we wish to show how to construct Lebesgue measure on the real line $\boldsymbol{R}$ from the length of an interval.

It is natural to define the **length** of the half-open interval $(a, b]$ to be the real number $b - a$ and the length of the sets $(-\infty, b] = \{x \in \boldsymbol{R} : x \leqslant b\}$, and $(a, +\infty) = \{x \in \boldsymbol{R} : a < x\}$, and $(-\infty, +\infty)$ to be the extended real number $+\infty$. We define the length of the union of a finite number of disjoint sets of these forms to be the sum of the corresponding lengths. Thus, the length of

$$\bigcup_{j=1}^{n} (a_j, b_j] \quad \text{is} \quad \sum_{j=1}^{n} (b_j - a_j)$$

provided the intervals do not intersect.

At first glance one might think that we have defined a measure on the family $\boldsymbol{F}$ of all sets which are finite unions of sets of the form

$$(9.1) \qquad (a, b], \quad (-\infty, b], \quad (a, +\infty), \quad (-\infty, +\infty).$$

However, this is not the case since the countable union of sets in $\boldsymbol{F}$ is not necessarily in $\boldsymbol{F}$, so that $\boldsymbol{F}$ is not a σ-algebra in the sense of Definition 2.1.

9.1 DEFINITION. A family $\boldsymbol{A}$ of subsets of a set X is said to be an **algebra** or a **field** in case:

(i) $\emptyset$, X belong to $\boldsymbol{A}$.

(ii) If E belongs to $\boldsymbol{A}$, then its complement $X \setminus E$ also belongs to $\boldsymbol{A}$.

(iii) If $E_1, \ldots, E_n$ belong to $\boldsymbol{A}$, then their union $\bigcup_{j=1}^{n} E_j$ also belongs to $\boldsymbol{A}$.

It is convenient to define the notion of a measure on an algebra. In doing so, we require the set function to be countably additive over sequences whose union belongs to the algebra.

9.2 DEFINITION. If $\boldsymbol{A}$ is an algebra of subsets of a set X, then a **measure** on $\boldsymbol{A}$ is an extended real-valued function μ defined on $\boldsymbol{A}$ such that (i) $\mu(\emptyset) = 0$, (ii) $\mu(E) \geqslant 0$ for all $E \in \boldsymbol{A}$, and (iii) if (E_n) is any disjoint sequence of sets in $\boldsymbol{A}$ such that $\bigcup_{n=1}^{\infty} E_n$ belongs to $\boldsymbol{A}$, then

$$\mu\left(\bigcup_{n=1}^{\infty} E_n\right) = \sum_{n=1}^{\infty} \mu(E_n).$$

It seems reasonably clear, but not entirely obvious, that length gives a measure. We now prove this fact.

9.3 LEMMA. *The collection $\boldsymbol{F}$ of all finite unions of sets of the form* (9.1) *is an algebra of subsets of $\boldsymbol{R}$ and length is a measure on $\boldsymbol{F}$.*

PROOF. It is readily seen that $\boldsymbol{F}$ is an algebra. If l denotes the length function, then conditions 9.2(i) and (ii) are trivial. To prove (iii) it is enough to show that if one of the sets of the form (9.1) is the union of a countable collection of sets of this form, then the length adds up correctly. We shall treat an interval of the form $(a, b]$, leaving the other possibilities as exercises. Suppose, then, that

$$(a, b] = \bigcup_{j=1}^{\infty} (a_j, b_j], \tag{9.2}$$

where the intervals $(a_j, b_j]$ are disjoint. Let $(a_1, b_1], \ldots, (a_n, b_n]$ be any finite collection of such intervals and suppose that

$$a \leqslant a_1 < b_1 \leqslant a_2 < \cdots < b_{n-1} \leqslant a_n < b_n \leqslant b.$$

(This may require a renumbering of the indices, but it can always be arranged.) Now

$$\begin{aligned} \sum_{j=1}^{n} l\big((a_j, b_j]\big) &= \sum_{j=1}^{n} (b_j - a_j) \\ &\leqslant b_n - a_1 \leqslant b - a = l\big((a, b]\big). \end{aligned}$$

Since n is arbitrary, we infer that

$$\sum_{j=1}^{\infty} l\big((a_j, b_j]\big) \leqslant l\big((a, b]\big). \tag{9.3}$$

Conversely, let $\varepsilon > 0$ be arbitrary, and let (ε_j) be a sequence of positive numbers with $\sum \varepsilon_j < \varepsilon$. By renumbering, if necessary, we may suppose that $a_1 = a$. Now consider the open intervals

$$\begin{aligned} I_1 &= (a_1 - \varepsilon_1, b_1 + \varepsilon_1), \\ I_j &= (a_j, b_j + \varepsilon_j), \qquad j \geqslant 2. \end{aligned}$$

In view of (9.2) it follows that the open sets $\{I_j : j \in N\}$ form a covering of the compact interval $[a, b]$. Therefore, this interval is covered by a finite number of the intervals, say by the intervals $I_1, I_2, \ldots, I_m$. By renumbering and discarding some extra intervals we may assume that

$$a = a_1 \leqslant a_2 < b_1 + \varepsilon_1 < \cdots < a_m < b_{m-1} + \varepsilon_{m-1} \leqslant b < b_m + \varepsilon_m.$$

It follows from this chain of inequalities that

$$\begin{aligned} b - a \leqslant (b_m + \varepsilon_m) - a_1 &\leqslant \sum_{j=1}^{m} [(b_j + \varepsilon_j) - a_j] \\ &< \sum_{j=1}^{m} (b_j - a_j) + \varepsilon \leqslant \sum_{j=1}^{\infty} (b_j - a_j) + \varepsilon. \end{aligned}$$

Since ε is arbitrary, we infer that

$$l\big((a, b]\big) \leqslant \sum_{j=1}^{\infty} l\big((a_j, b_j]\big).$$

Combining this inequality with (9.3), we conclude that the length function l is countably additive on $\boldsymbol{F}$. Q.E.D.

THE EXTENSION OF MEASURES

Now that we have given a significant example of a measure defined on an algebra of sets, we return to the general situation. We shall show that if $\boldsymbol{A}$ is any algebra of subsets of a set X and if μ is a measure defined on $\boldsymbol{A}$, then there exists a σ-algebra $\boldsymbol{A}^*$ containing $\boldsymbol{A}$ and a measure μ^* defined on $\boldsymbol{A}^*$ such that $\mu^*(E) = \mu(E)$ for E in $\boldsymbol{A}$. In other words, the measure μ can be extended to a measure on a σ-algebra $\boldsymbol{A}^*$ of subsets of X which contains $\boldsymbol{A}$. The procedure that we employ is the following: we shall use μ to obtain a function defined for *all* subsets of X, and then pick out a collection of sets on which a certain additivity property holds.

9.4 DEFINITION. If B is an arbitrary subset of X, we define

$$\mu^*(B) = \inf \sum_{j=1}^{\infty} \mu(E_j), \tag{9.4}$$

where the infimum is extended over all sequences (E_j) of sets in $\boldsymbol{A}$ such that

$$B \subseteq \bigcup_{j=1}^{\infty} E_j. \tag{9.5}$$

It should be remarked that the function μ^* just defined is usually called the **outer measure** generated by μ. Although this terminology is unfortunate because μ^* is not generally a measure, μ^* does have a few properties reminiscent of a measure.

9.5 LEMMA. *The function μ^* of Definition 9.4 satisfies the following:*

(a) $\mu^*(\emptyset) = 0$.
(b) $\mu^*(B) \geqslant 0$, *for* $B \subseteq X$.
(c) *If* $A \subseteq B$, *then* $\mu^*(A) \leqslant \mu^*(B)$.
(d) *If* $B \in \boldsymbol{A}$, *then* $\mu^*(B) = \mu(B)$.
(e) *If* (B_n) *is a sequence of subsets of* X, *then*

$$\mu^*\left(\bigcup_{n=1}^{\infty} B_n\right) \leqslant \sum_{n=1}^{\infty} \mu^*(B_n).$$

PROOF. Statements (a), (b), and (c) are immediate consequences of the Definition 9.4.

(d) Since $\{B, \emptyset, \emptyset, \ldots\}$ is a countable collection of sets in $\boldsymbol{A}$ whose union contains B, it follows that

$$\mu^*(B) \leqslant \mu(B) + 0 + 0 + \cdots = \mu(B).$$

Conversely, if (E_n) is any sequence from $\boldsymbol{A}$ with $B \subseteq \bigcup E_n$, then

$B = \bigcup (B \cap E_n)$. Since μ is a measure on $\boldsymbol{A}$, then

$$\mu(B) \leqslant \sum_{n=1}^{\infty} \mu(B \cap E_n) \leqslant \sum_{n=1}^{\infty} \mu(E_n),$$

from which it follows that $\mu(B) \leqslant \mu^*(B)$.

To establish (e), let $\varepsilon > 0$ be arbitrary and for each n choose a sequence (E_{nk}) of sets in $\boldsymbol{A}$ such that

$$B_n \subseteq \bigcup_{k=1}^{\infty} E_{nk} \quad \text{and} \quad \sum_{k=1}^{\infty} \mu(E_{nk}) \leqslant \mu^*(B_n) + \frac{\varepsilon}{2^n} \cdot$$

Since $\{E_{nk} : n, k \in \boldsymbol{N}\}$ is a countable collection from $\boldsymbol{A}$ whose union contains $\bigcup B_n$, it follows from the definition of μ^* that

$$\mu^*\left(\bigcup_{n=1}^{\infty} B_n\right) \leqslant \sum_{n=1}^{\infty} \sum_{k=1}^{\infty} \mu(E_{nk}) \leqslant \sum_{n=1}^{\infty} \mu^*(B_n) + \varepsilon.$$

Since ε is arbitrary, the desired inequality is obtained. Q.E.D.

Property (e) of Lemma 9.5 is referred to by saying that μ^* is **countably subadditive.**

Although μ^* has the advantage that it is defined for arbitrary subsets of X, it has the defect that it is not necessarily countably (or even finitely) additive. We are willing to restrict μ^* to a smaller σ-algebra provided we can find one containing $\boldsymbol{A}$ and over which μ^* has the property of countable additivity. There is a remarkable condition due to Carathéodory which provides the desired restriction of the domain of μ^*.

9.6 DEFINITION. A subset E of X is said to be **μ^*-measurable** if

$$\mu^*(A) = \mu^*(A \cap E) + \mu^*(A \setminus E) \tag{9.6}$$

for all subsets A of X. The collection of all μ^*-measurable sets is denoted by $\boldsymbol{A}^*$.

Condition (9.6) indicates an additivity property on μ^*. In loose terms, a set E is μ^*-measurable in case it and its complement are sufficiently separated that they divide an arbitrary set A additively.

9.7 CARATHÉODORY EXTENSION THEOREM. *The collection A^* of all μ^*-measurable sets is a σ-algebra containing A. Moreover, if (E_n) is a disjoint sequence in A^*, then*

$$\mu^*\Big(\bigcup_{n=1}^{\infty} E_n\Big) = \sum_{n=1}^{\infty} \mu^*(E_n). \tag{9.7}$$

PROOF. It is clear that $\emptyset$ and X are μ^*-measurable, and that if $E \in A^*$, then its complement $X \setminus E$ belongs to A^*.

Next we shall show that A^* is closed under intersections. Indeed, suppose that E and F are μ^*-measurable. Then for any $A \subseteq X$ and $E \in A^*$, we have

$$\mu^*(A \cap F) = \mu^*(A \cap F \cap E) + \mu^*\big((A \cap F) \setminus E\big)$$

Since $F \in A^*$, then

$$\mu^*(A) = \mu^*(A \cap F) + \mu^*(A \setminus F).$$

Let $B = A \setminus (E \cap F)$, then it is readily seen that $B \cap F = (A \cap F) \setminus E$ and $B \setminus F = A \setminus F$; since $F \in A^*$ it follows that

$$\mu^*\big(A \setminus (E \cap F)\big) = \mu^*\big((A \cap F) \setminus E\big) + \mu^*(A \setminus F).$$

Combining these three relations, we obtain

$$\mu^*(A) = \mu^*(A \cap E \cap F) + \mu^*\big(A \setminus (E \cap F)\big),$$

which shows that $E \cap F$ belongs to A^*. Since A^* is closed under intersection and complementation, it follows that A^* is an algebra.

Suppose that $E, F \in A^*$ and that $E \cap F = \emptyset$. If we take A to be $A \cap (E \cup F)$ in (9.6), we obtain

$$\mu^*\big(A \cap (E \cup F)\big) = \mu^*(A \cap E) + \mu^*(A \cap F).$$

For $A = X$, this relation implies that μ^* is additive on A^*.

We shall now show that A^* is a σ-algebra and that μ^* is countably additive on A^*. Let (E_k) be a disjoint sequence in A^* and let $E = \bigcup E_k$. From the preceding paragraph, we know that $F_n = \bigcup_{k=1}^{n} E_k$ belongs to A^*, and that if A is any subset of X, then

$$\mu^*(A) = \mu^*(A \cap F_n) + \mu^*(A \setminus F_n) = \sum_{k=1}^{n} \mu^*(A \cap E_k) + \mu^*(A \setminus F_n).$$

Since $F_n \subseteq E$, then $A \setminus E \subseteq A \setminus F_n$ and letting $n \to \infty$ the above relations yields

$$\sum_{k=1}^{\infty} \mu^*(A \cap E_k) + \mu^*(A \setminus E) \leqslant \mu^*(A).$$

On the other hand, it follows from Lemma 9.5(e) that

$$\mu^*(A \cap E) \leqslant \sum_{k=1}^{\infty} \mu^*(A \cap E_k),$$

$$\mu^*(A) \leqslant \mu^*(A \cap E) + \mu^*(A \setminus E).$$

On combining the last three inequalities we infer that

$$\mu^*(A) = \mu^*(A \cap E) + \mu^*(A \setminus E) = \sum_{k=1}^{\infty} \mu^*(A \cap E_k) + \mu^*(A \setminus E).$$

In particular, this shows that $E = \bigcup_{k=1}^{\infty} E_k$ is μ^*-measurable. On taking $A = E$, we obtain (9.7).

It remains to prove that $\mathbf{A} \subseteq \mathbf{A}^*$. It was proved in Lemma 9.5(d) that if $E \in \mathbf{A}$, then $\mu^*(E) = \mu(E)$, but we need to show that E is μ^*-measurable. Let A be an arbitrary subset of X; it follows from Lemma 9.5(e) that

$$\mu^*(A) \leqslant \mu^*(A \cap E) + \mu^*(A \setminus E).$$

To establish the opposite inequality, let $\varepsilon > 0$ be arbitrary and let (F_n) be a sequence in $\mathbf{A}$ such that $A \subseteq \bigcup F_n$ and

$$\sum_{n=1}^{\infty} \mu(F_n) \leqslant \mu^*(A) + \varepsilon.$$

Since $A \cap E \subseteq \bigcup (F_n \cap E)$ and $A \setminus E \subseteq \bigcup (F_n \setminus E)$, it follows from Lemma 9.5(e) that

$$\mu^*(A \cap E) \leqslant \sum_{n=1}^{\infty} \mu(F_n \cap E), \qquad \mu^*(A \setminus E) \leqslant \sum_{n=1}^{\infty} \mu(F_n \setminus E).$$

Hence we have

$$\begin{aligned} \mu^*(A \cap E) + \mu^*(A \setminus E) &\leqslant \sum_{n=1}^{\infty} \{\mu(F_n \cap E) + \mu(F_n \setminus E)\} \\ &= \sum_{n=1}^{\infty} \mu(F_n) \leqslant \mu^*(A) + \varepsilon. \end{aligned}$$

Since ε is arbitrary, the desired inequality is established and the set E belongs to A^*. Q.E.D.

The Carathéodory Extension Theorem shows that a measure μ on an algebra A can always be extended to a measure μ^* on a σ-algebra A^* containing A. The σ-algebra A^* obtained in this way is automatically **complete** in the sense that if $E \in A^*$ with $\mu^*(E) = 0$, and if $B \subseteq E$, then $B \in A^*$ and $\mu^*(B) = 0$. To prove this, let A be an arbitrary subset of X and employ Lemma 9.5(c) to observe that

$$\mu^*(A) = \mu^*(E) + \mu^*(A) \geqslant \mu^*(A \cap B) + \mu^*(A \setminus B);$$

and, as before, the inequality

$$\mu^*(A) \leqslant \mu^*(A \cap B) + \mu^*(A \setminus B)$$

follows from Lemma 9.5(e). Hence B is μ^*-measurable and

$$0 \leqslant \mu^*(B) \leqslant \mu^*(E) \leqslant 0.$$

We shall now show that in the case that μ is a σ-finite measure, it has a unique extension to a measure on A^*.

9.8 Hahn Extension Theorem. *Suppose that μ is a σ-finite measure on an algebra A. Then there exists a unique extension of μ to a measure on A^*.*

PROOF. The fact that μ^* gives a measure on A^* was proved in Theorem 9.7 even without the σ-finiteness assumption. To establish the uniqueness, let ν be a measure on A^* which agrees with μ on A.

First suppose that μ and therefore both μ^* and ν are finite measures. Let E be any set in A^* and let (E_n) be a sequence in A such that $E \subseteq \bigcup E_n$. Since ν is a measure and agrees with μ on A we have

$$\nu(E) \leqslant \nu\left(\bigcup_{n=1}^{\infty} E_n\right) \leqslant \sum_{n=1}^{\infty} \nu(E_n) = \sum_{n=1}^{\infty} \mu(E_n).$$

Therefore $\nu(E) \leqslant \mu^*(E)$ for any $E \in A^*$. Since μ^* and ν are additive, $\mu^*(E) + \mu^*(X \setminus E) = \nu(E) + \nu(X \setminus E)$. Since the terms on the right-hand side are finite and not greater than the corresponding terms on the left side, we infer that $\mu^*(E) = \nu(E)$ for all $E \in A^*$. This establishes the uniqueness when μ is a finite measure.

Suppose that μ is σ-finite and let (F_n) be an increasing sequence of sets in $\boldsymbol{A}$ with $\mu(F_n) < +\infty$ and $X = \bigcup F_n$. From the preceding paragraph, $\mu^*(E \cap F_n) = \nu(E \cap F_n)$ for each E in $\boldsymbol{A}^*$. Therefore

$$\begin{aligned}\mu^*(E) &= \lim \mu^*(E \cap F_n)\\ &= \lim \nu(E \cap F_n) = \nu(E),\end{aligned}$$

so that μ^* and ν agree on $\boldsymbol{A}^*$. Q.E.D.

LEBESGUE MEASURE

We now return to the considerations that prompted the foregoing extension procedure, namely, to the generation of a measure on the real line $\boldsymbol{R}$. In Lemma 9.3 we saw that the set $\boldsymbol{F}$ of all finite unions of sets of the form

$$(a, b], \quad (-\infty, b], \quad (a, +\infty), \quad (-\infty, +\infty),$$

was an algebra of subsets of $\boldsymbol{R}$ and that the length function l gives a measure on this algebra $\boldsymbol{F}$. If we apply the extension procedure to l and $\boldsymbol{F}$, we generate a measure space $(\boldsymbol{R}, \boldsymbol{F}^*, l^*)$. The σ-algebra $\boldsymbol{F}^*$ obtained in this construction is called the collection of **Lebesgue measurable sets** and the measure l^* on $\boldsymbol{F}^*$ is called **Lebesgue measure.**†

Although we sometimes wish to work with $(\boldsymbol{R}, \boldsymbol{F}^*, l^*)$, it is often more convenient to deal with the smallest σ-algebra containing $\boldsymbol{F}$ than with all of $\boldsymbol{F}^*$. It is readily seen that this smallest σ-algebra is exactly the collection of Borel sets. The restriction of Lebesgue measure to the Borel sets is called either **Borel** or **Lebesgue measure**. Lest the reader feel that restricting to $\boldsymbol{B}$ weakens the theory by substantially lessening the collection of measurable sets and functions, we call attention to Exercise 9.K where it is seen that every Lebesgue measurable set is contained in a Borel measurable set with the same measure, and every Lebesgue measurable function is almost everywhere equal to a Borel measurable function.

† It might be thought that every subset of the real line is Lebesgue measurable, but this is not the case. For the construction of sets which are not Lebesgue measurable, consult Reference [7], pp. 67–70.

Sometimes it is more convenient to use a notion of the magnitude of an interval other than length. This can be treated as follows. Let g be a monotone increasing function on $\boldsymbol{R}$ to $\boldsymbol{R}$ so that $x \leqslant y$ implies that $g(x) \leqslant g(y)$. In addition, we shall assume that g is continuous on the right at every point, so that

$$g(c) = \lim_{h \to 0+} g(c + h).$$

Since g is monotone, it also follows that

$$\lim_{x \to -\infty} g(x), \qquad \lim_{x \to +\infty} g(x)$$

both exist, although they may be $-\infty$ or $+\infty$.

For such a function we define

$$\begin{aligned} \mu_g\big((a, b]\big) &= g(b) - g(a), \\ \mu_g\big((-\infty, b]\big) &= g(b) - \lim_{x \to -\infty} g(x), \\ \mu_g\big((a, +\infty)\big) &= \lim_{x \to +\infty} g(x) - g(a), \\ \mu_g\big((-\infty, \infty)\big) &= \lim_{x \to +\infty} g(x) - \lim_{x \to -\infty} g(x). \end{aligned}$$

We further define μ_g on the algebra $\boldsymbol{F}$ of finite disjoint unions of such sets to be the corresponding sums. If the reader will check the details of the proof of Lemma 9.3, he will see that it can be easily modified to show that μ_g gives a σ-finite measure on the algebra $\boldsymbol{F}$. Therefore, this measure has a unique extension, which we also denote by μ_g to the algebra of all Borel subsets of $\boldsymbol{R}$. This extension is often referred to as the **Borel-Stieltjes measure** generated by g. (Of course, by applying Theorem 9.7, μ_g has an extension to a complete σ-algebra which contains the Borel sets. This extension is called the **Lebesgue-Stieltjes** measure generated by g.)

LINEAR FUNCTIONALS ON *C*

We shall conclude this chapter by showing that there is an intimate correspondence between Borel-Stieltjes measures on a finite closed interval $J = [a, b]$ and bounded positive linear functionals on the

Banach space $C(J)$ of all continuous functions on J to $\boldsymbol{R}$ with the norm

$$\|f\| = \sup \{|f(x)| : x \in J\}. \tag{9.8}$$

This result, due to F. Riesz, has been considerably extended in many directions. Indeed, it is taken as the point of departure for the development of a theory of integration by many authors who prefer to regard the integral as a linear functional on spaces of continuous functions. We choose to take a very concrete approach to this theorem and offer a proof which is closely parallel to the Riemann–Stieltjes integral version presented in Reference [1], pp. 290–294.

9.9 RIESZ REPRESENTATION THEOREM. *If G is a bounded positive linear functional on $C(J)$, then there exists a measure γ defined on the Borel subsets of $\boldsymbol{R}$ such that*

$$G(f) = \int_J f \, d\gamma \tag{9.9}$$

for all f in $C(J)$. Moreover, the norm $\|G\|$ of G equals $\gamma(J)$.

PROOF. If t is such that $a \leqslant t < b$ and n is a sufficiently large natural number, let $\varphi_{t,n}$ be the function in $C(J)$ which equals 1 on $[a, t]$, which equals 0 on $(t + 1/n, b]$, and which is linear on $(t, t + 1/n]$. If $n \leqslant m$ and $x \in J$, then $0 \leqslant \varphi_{t,m}(x) \leqslant \varphi_{t,n}(x) \leqslant 1$, so that the real sequence $\big(G(\varphi_{t,n})\big)$ is bounded and decreasing. If $t \in [a, b)$, we define

$$g(t) = \lim_{n \to +\infty} G(\varphi_{t,n}).$$

Further, set $g(t) = 0$ for $t < a$; if $t \geqslant b$, we set $g(t) = G(\varphi_b)$ where $\varphi_b(x) = 1$ for all $x \in J$. It is readily seen that g is a monotone increasing function on $\boldsymbol{R}$.

We claim that g is continuous from the right. This is clear if $t < a$ or $t \geqslant b$. Suppose that $t \in [a, b)$ and $\varepsilon > 0$ and let

$$n > \sup \{2, \|G\|\varepsilon^{-1}\}$$

be so large that

$$g(t) \leqslant G(\varphi_{t,n}) \leqslant g(t) + \varepsilon.$$

If ψ_n is the function in $C(J)$ which equals 1 on $[a, t + n^{-2}]$, which equals 0 on $(t + n^{-1} - n^{-2}, b]$, and which is linear on

$$(t + n^{-2}, t + n^{-1} - n^{-2}],$$

then an exercise in analytic geometry shows that $\|\psi_n - \varphi_{t,n}\| \leqslant 1/n$. Therefore

$$G(\psi_n) \leqslant G(\varphi_{t,n}) + \left(\frac{1}{n}\right) \|G\| \leqslant g(t) + 2\varepsilon,$$

so that $g(t) \leqslant g(t + n^{-2}) \leqslant g(t) + 2\varepsilon$.

According to the Hahn Extension Theorem there exists a measure γ on the Borel subsets of $\boldsymbol{R}$ such that $\gamma\big((\alpha, \beta]\big) = g(\beta) - g(\alpha)$. In particular, this shows that $\gamma(E) = 0$, if $E \cap J = \emptyset$, that

$$\gamma\big([a, c]\big) = \gamma\big((a - 1, c]\big) = g(c),$$

and that $\|G\| = \|G(\varphi_b)\| = g(b) = \gamma(J)$.

It remains to show that equation (9.9) holds for f in $C(J)$. If $\varepsilon > 0$, since f is uniformly continuous on J, there is a $\delta(\varepsilon) > 0$ such that if $|x - y| < \delta(\varepsilon)$ and $x, y \in J$, then $|f(x) - f(y)| < \varepsilon$. Now let $a = t_0 < t_1 < \cdots < t_m = b$ be such that $\sup\{t_k - t_{k-1}\} < \frac{1}{2}\delta(\varepsilon)$ and choose n so large that $2/n < \inf\{t_k - t_{k-1}\}$ and that for $k = 1, \ldots, m$, then

$$g(t_k) \leqslant G(\varphi_{t_k, n}) \leqslant g(t_k) + \varepsilon(m\|f\|)^{-1}. \tag{9.10}$$

We now consider functions defined on J by

$$f_1(x) = f(t_1)\,\varphi_{t_1, n}(x) + \sum_{k=2}^{m} f(t_k)\{\varphi_{t_k, n}(x) - \varphi_{t_{k-1}, n}(x)\},$$

$$f_2(x) = f(t_1)\,\chi_{[t_0, t_1]}(x) + \sum_{k=2}^{m} f(t_k)\,\chi_{(t_{k-1}, t_k]}(x).$$

Note that $f_1 \in C(J)$ and that f_2 is a step function on J. It is clear that $\sup\{|f_2(x) - f(x)| : x \in J\} \leqslant \varepsilon$ and as an exercise (or see [1], p. 292) the reader can show that $\|f_1 - f\| \leqslant \varepsilon$. Therefore we have

$$|G(f) - G(f_1)| \leqslant \varepsilon\,\|G\|.$$

In view of (9.10) we see that if $2 \leqslant k \leqslant m$, then

$$|G(\varphi_{t_k,n} - \varphi_{t_{k-1},n}) - \{g(t_k) - g(t_{k-1})\}| \leqslant \varepsilon(m\,\|f\|)^{-1}.$$

Apply G to f_1 and integrate f_2 with respect to γ. The inequality just obtained yields

$$|G(f_1) - \int_J f_2\, d\gamma| \leqslant \varepsilon.$$

But since f_2 lies within ε of f, we also have

$$\left| \int_J f_2\, d\gamma - \int_J f\, d\gamma \right| \leqslant \varepsilon\, \gamma(J).$$

Combining the inequalities, we arrive at the inequality

$$|G(f) - \int_J f\, d\gamma| \leqslant \varepsilon\,(2\|G\| + 1),$$

and since ε is arbitrary, we deduce (9.9). Q.E.D.

If the reader will check the proof of Lemma 8.13, he will see that an arbitrary bounded linear functional G on $C(J)$ can be written as the difference $G^+ - G^-$ of two positive bounded linear functionals. Making use of this observation, one can extend the Riesz Representation Theorem given above to represent a bounded linear functional on $C(J)$ by means of integration with respect to a charge defined on the Borel subsets of J.

EXERCISES

9.A. Establish that the family $\boldsymbol{F}$ of all finite unions of sets of the form (9.1) is an algebra of sets in $\boldsymbol{R}$.

9.B. Show that the family $\boldsymbol{G}$ of all finite unions of sets of the form

$$(a, b), \quad (-\infty, b), \quad (a, +\infty), \quad (-\infty, +\infty)$$

is *not* an algebra of sets in $\boldsymbol{R}$. However, the σ-algebra generated by $\boldsymbol{G}$ is the family of Borel sets.

9.C. Show that if the set $(a, +\infty)$ is the union of a disjoint sequence of sets $(a_n, b_n]$, then

$$\sum_{n=1}^{\infty} l((a_n, b_n]) = +\infty.$$

9.D. Let X be the set of all rational numbers r satisfying $0 < r \leqslant 1$ and let $\boldsymbol{A}$ be the family of all finite unions of "half-open intervals" of the form $\{r \in X : a < r \leqslant b\}$, where $0 \leqslant a \leqslant b \leqslant 1$ and $a, b \in X$. Show that $\boldsymbol{A}$ is an algebra of subsets of X. Moreover, every non-empty set in $\boldsymbol{A}$ is infinite. However, the σ-algebra generated by $\boldsymbol{A}$ consists of all subsets of X.

9.E. If E is a countable subset of $\boldsymbol{R}$, then it has Lebesgue measure zero.

9.F. Let $I_n = (n, n + 1]$, for $n = 0, \pm 1, \pm 2, \ldots$. If a subset E is contained in the union of a finite number of the $\{I_n\}$, then $l^*(E) < +\infty$. However, construct a Lebesgue measurable set E with $l^*(E) < +\infty$ such that $l^*(E \cap I_n) > 0$ for all n. Show that a subset E of $\boldsymbol{R}$ is Lebesgue measurable if and only if $E \cap I_n$ is Lebesgue measurable for each n.

9.G. If A is a Lebesgue measurable subset of $\boldsymbol{R}$ and $\varepsilon > 0$, show that there exists an open set $G_\varepsilon \supseteq A$ such that

$$l^*(A) \leqslant l^*(G_\varepsilon) \leqslant l^*(A) + \varepsilon.$$

9.H. If B is a Lebesgue measurable subset of $\boldsymbol{R}$, if $\varepsilon > 0$, and if $B \subseteq I_n = (n, n + 1]$, then there exists a compact set $K_\varepsilon \subseteq B$ such that

$$l^*(K_\varepsilon) \leqslant l^*(B) \leqslant l^*(K_\varepsilon) + \varepsilon.$$

(*Hint:* Apply the Exercise 9.G to $A = I_n \setminus B$.)

9.I. If A is an arbitrary Lebesgue measurable set in $\boldsymbol{R}$, apply the preceding exercises to show that

$$l^*(A) = \inf\{l^*(G) : A \subseteq G, G \text{ open}\},$$
$$l^*(A) = \sup\{l^*(K) : K \subseteq A, K \text{ compact}\}.$$

9.J. Let $\lambda = l^*$ denote Lebesgue measure on $\boldsymbol{R}$, and let A be a Lebesgue measurable set with $\lambda(A) < +\infty$. If $\varepsilon > 0$, there exists an open set which is the union of a finite number of open intervals such that

$$\|\chi_A - \chi_G\|_1 = |\lambda(A) - \lambda(G)| < \varepsilon.$$

Moreover, if $\varepsilon > 0$ there exists a continuous function f such that

$$\|\chi_A - f\|_1 = \int |\chi_A - f| \, d\lambda < \varepsilon.$$

9.K. Let A be a Lebesgue measurable subset of $\boldsymbol{R}$. Show that there exists a Borel measurable subset B of $\boldsymbol{R}$ such that $A \subseteq B$ and such that $l^*(B \setminus A) = 0$. (*Hint:* Consider the case where $l^*(A) < +\infty$ first.) Show that every Lebesgue measurable set is the union of a Borel measurable set (with the same measure) and a set of Lebesgue measure zero. In the terminology of Exercise 3.L, this asserts that the Lebesgue algebra is the **completion** of the Borel algebra. As a consequence of Exercise 3.N, we infer that every Lebesgue measurable function is almost everywhere equal to a Borel measurable function.

9.L. If g belongs to $L(\boldsymbol{R}, \boldsymbol{B}, \lambda)$ and $\varepsilon > 0$, then there exists a continuous function f such that

$$\|g - f\|_1 = \int |g - f| \, d\lambda < \varepsilon.$$

9.M. If $\boldsymbol{B}$ is the Borel algebra and λ is Lebesgue measure on $\boldsymbol{B}$, show that (i) $\lambda(G) > 0$ for every open set G, (ii) $\lambda(K) < +\infty$ for every compact set K, and (iii) $\lambda(x + E) = \lambda(E)$ for all $E \in \boldsymbol{B}$. (Here $x + E = \{x + y : y \in E\}$.)

9.N. Let X be a set, $\boldsymbol{A}$ an algebra of subsets of X, and μ a measure on $\boldsymbol{A}$. If B is an arbitrary subset of X, let $\mu'(B)$ be defined to be

$$\mu'(B) = \inf\{\mu(A) : B \subseteq A \in \boldsymbol{A}\}.$$

Show that $\mu'(E) = \mu(E)$ for all $E \in \boldsymbol{A}$ and that $\mu^*(B) \leqslant \mu'(B)$. Moreover, $\mu^* = \mu'$ in case X is the countable union of sets with finite μ-measure. Is μ' countably subadditive in the sense of 9.5(e)?

9.O. Let X be an uncountable set and let $\boldsymbol{A}$ be the collection of sets E which are either finite or have finite complement. In the former case let $\mu(E) = 0$; in the latter, let $\mu(E) = +\infty$. Show that μ is a measure on $\boldsymbol{A}$. Calculate the outer measure μ^* corresponding to Definition 9.4. Calculate the set function μ' defined in Exercise 9.N. Are they the same?

9.P. Let X be a set and let α be defined for arbitrary subsets of X to $\boldsymbol{R}$ and satisfy

$$0 \leqslant \alpha(E) \leqslant \alpha(E \cup F) \leqslant \alpha(E) + \alpha(F),$$

when E and F are subsets of X. Let $\boldsymbol{S}$ be the collection of all subsets E of X such that

$$\alpha(A) = \alpha(A \cap E) + \alpha(A \setminus E)$$

for all $A \subseteq X$. If $\boldsymbol{S} \neq \emptyset$, it is an algebra and α is additive on $\boldsymbol{S}$.

9.Q. It may happen that the collection $\boldsymbol{S}$ in Exercise 9.P is empty. For example, let $\alpha(E) = 1$ for all $E \subseteq X$.

9.R. Let X and $\boldsymbol{A}$ be as in Exercise 9.D, and let $\boldsymbol{A}_1$ be the σ-algebra generated by $\boldsymbol{A}$. Let μ_1 be the counting measure on $\boldsymbol{A}_1$ and let $\mu_2 = 2\mu_1$. Show that $\mu_1 = \mu_2$ on $\boldsymbol{A}$ but not on $\boldsymbol{A}_1$. (Hence the σ-finiteness hypothesis in Theorem 9.8 cannot be dropped.)

9.S. Let g be a monotone increasing and right continuous function on $\boldsymbol{R}$ to $\boldsymbol{R}$. If μ_g is defined as at the end of this section, show that μ_g is a measure on the algebra $\boldsymbol{F}$.

9.T. Consider the following functions defined for $x \in \boldsymbol{R}$ by:

(a) $g_1(x) = 2x$, (b) $g_2(x) = \text{Arc} \tan x$,

(c) $g_3(x) = 0, x < 0,$
$\qquad = 1, x \geqslant 0,$

(d) $g_4(x) = 0, x < 0,$
$\qquad = x, x \geqslant 0.$

Describe the Borel-Stieltjes measures determined by these functions. Which of these measures are absolutely continuous with respect to Borel measure? What are their Radon-Nikodým derivatives? Which of these measures are singular with respect to Borel measure? Which of these measures are finite? With respect to which of these measures is Borel measure absolutely continuous?

9.U. Let μ be a finite measure on the Borel sets $\boldsymbol{B}$ of $\boldsymbol{R}$ and let $g(x) = \mu\big((-\infty, x]\big)$ for $x \in \boldsymbol{R}$. Show that g is monotone increasing and right continuous, and that

$$\mu\big((a, b]\big) = g(b) - g(a)$$

when $-\infty < a \leqslant b < +\infty$. Show that $\mu(\boldsymbol{R}) = \lim_{x \to \infty} g(x)$.

9.V. Let f be Riemann integrable on $[a, b]$ to $\boldsymbol{R}$. Then there exists a monotone increasing sequence (φ_n) and a monotone decreasing sequence (ψ_n) of step functions such that $\varphi_n(x) \leqslant f(x) \leqslant \psi_n(x)$ for $x \in [a, b]$ and

$$\lim \int \varphi_n \, d\lambda = \lim \int \psi_n \, d\lambda.$$

(Here λ denotes Lebesgue measure.) Show that $f = \lim \psi_n = \lim \varphi_n$ almost everywhere, that f is Lebesgue measurable, and that

$$\int f \, d\lambda = \int_a^b f(x) \, dx.$$

CHAPTER 10

Product Measures

Let X and Y be two sets; then the **Cartesian product** $Z = X \times Y$ is the set of all ordered pairs (x, y) with $x \in X$ and $y \in Y$. We shall first show that the Cartesian product of two measurable spaces $(X, \boldsymbol{X})$ and $(Y, \boldsymbol{Y})$ can be made into a measurable space in a natural fashion. Next we shall show that if measures are given on each of the factor spaces, we can define a measure on the product space. Finally, we shall relate integration with respect to the product measure and iterated integration with respect to the measures in the factor spaces. The model to be kept in mind throughout this discussion is the plane, which we regard as the product $\boldsymbol{R} \times \boldsymbol{R}$.

10.1 Definition. If $(X, \boldsymbol{X})$ and $(Y, \boldsymbol{Y})$ are measurable spaces, then a set of the form $A \times B$ with $A \in \boldsymbol{X}$ and $B \in \boldsymbol{Y}$ is called a **measurable rectangle**, or simply a **rectangle**, in $Z = X \times Y$. We shall denote the collection of all finite unions of rectangles by $\boldsymbol{Z}_0$.

It is an exercise to show that every set in $\boldsymbol{Z}_0$ can be expressed as a finite *disjoint* union of rectangles in Z (see Exercise 10.D).

10.2 Lemma. *The collection $\boldsymbol{Z}_0$ is an algebra of subsets of Z.*

Proof. It is clear that the union of a finite number of sets in $\boldsymbol{Z}_0$ also belongs to $\boldsymbol{Z}_0$. Similarly, it follows from the first part of Exercise 10.E that the complement of a rectangle in Z belongs to $\boldsymbol{Z}_0$. Apply De Morgan's laws to see that the complement of any set in $\boldsymbol{Z}_0$ belongs to $\boldsymbol{Z}_0$. Q.E.D.

10.3 DEFINITION. If $(X, \mathbf{X})$ and $(Y, \mathbf{Y})$ are measurable spaces, then $\mathbf{Z} = \mathbf{X} \times \mathbf{Y}$ denotes the σ-algebra of subsets of $Z = X \times Y$ generated by rectangles $A \times B$ with $A \in \mathbf{X}$ and $B \in \mathbf{Y}$. We shall refer to a set in $\mathbf{Z}$ as a **$\mathbf{Z}$-measurable** set, or as a **measurable** subset of Z.

If $(X, \mathbf{X}, \mu)$ and $(Y, \mathbf{Y}, \nu)$ are measure spaces, it is natural to attempt to define a measure π on the subsets of $\mathbf{Z} = \mathbf{X} \times \mathbf{Y}$ which is the "product" of μ and ν in the sense that

$$\pi(A \times B) = \mu(A)\, \nu(B), \quad A \in \mathbf{X}, B \in \mathbf{Y}.$$

(Recall the convention that $0(\pm\infty) = 0$.) We shall now show that this can always be done.

10.4 PRODUCT MEASURE THEOREM. *If $(X, \mathbf{X}, \mu)$ and $(Y, \mathbf{Y}, \nu)$ are measure spaces, then there exists a measure π defined on $\mathbf{Z} = \mathbf{X} \times \mathbf{Y}$ such that*

$$\pi(A \times B) = \mu(A)\, \nu(B) \tag{10.1}$$

for all $A \in \mathbf{X}$ and $B \in \mathbf{Y}$. If these measure spaces are σ-finite, then there is a unique measure π with property (10.1).

PROOF. Suppose that the rectangle $A \times B$ is the disjoint union of a sequence $(A_j \times B_j)$ of rectangles; thus

$$\chi_{A \times B}(x, y) = \chi_A(x)\, \chi_B(y) = \sum_{j=1}^{\infty} \chi_{A_j}(x)\, \chi_{B_j}(y)$$

for all $x \in X$, $y \in Y$. Hold x fixed, integrate with respect to ν, and apply the Monotone Convergence Theorem to obtain

$$\chi_A(x)\, \nu(B) = \sum_{j=1}^{\infty} \chi_{A_j}(x)\, \nu(B_j).$$

A further application of the Monotone Convergence Theorem yields

$$\mu(A)\, \nu(B) = \sum_{j=1}^{\infty} \mu(A_j)\, \nu(B_j).$$

Now let $E \in \mathbf{Z}_0$; without loss of generality we may assume that

$$E = \bigcup_{j=1}^{n} (A_j \times B_j),$$

where the sets $A_j \times B_j$ are mutually disjoint rectangles. If we define $\pi_0(E)$ by

$$\pi_0(E) = \sum_{j=1}^{n} \mu(A_j)\, \nu(B_j),$$

the argument in the previous paragraph implies that π_0 is well-defined and countably additive on $\boldsymbol{Z}_0$. By Theorem 9.7, there is an extension of π_0 to a measure π on the σ-algebra $\boldsymbol{Z}$ generated by $\boldsymbol{Z}_0$. Since π is an extension of π_0, it is clear that (10.1) holds.

If $(X, \boldsymbol{X}, \mu)$ and $(Y, \boldsymbol{Y}, \nu)$ are σ-finite, then π_0 is a σ-finite measure on the algebra $\boldsymbol{Z}_0$ and the uniqueness of a measure satisfying (10.1) follows from the uniqueness assertion of the Hahn Extension Theorem 9.8. Q.E.D.

Theorem 10.4 establishes the existence of a measure π on the σ-algebra $\boldsymbol{Z}$ generated by the rectangles $\{A \times B : A \in \boldsymbol{X}, B \in \boldsymbol{Y}\}$ and such that (10.1) holds. Any such measure will be called a **product** of μ and ν. If μ and ν are both σ-finite, then they have a unique product. In the general case the extension procedure discussed in the previous section leads to a uniquely determined product measure. However, it will be seen in Exercise 10.S that it is possible for two distinct measures on $\boldsymbol{Z}$ to satisfy (10.1) if μ and ν are not σ-finite.

In order to relate integration with respect to a product measure and iterated integration, the notion of a section is useful.

10.5 Definition. If E is a subset of $Z = X \times Y$ and $x \in X$, then the ***x*-section of** E is the set

$$E_x = \{y \in Y : (x, y) \in E\}$$

Similarly, if $y \in Y$, then the ***y*-section** of E is the set

$$E^y = \{x \in X : (x, y) \in E\}.$$

If f is a function defined on Z to $\bar{\boldsymbol{R}}$, and $x \in X$, then the ***x*-section** of f is the function f_x defined on Y by

$$f_x(y) = f(x, y), \qquad y \in Y.$$

Similarly, if $y \in Y$, then the **y-section** of f is the function f^y defined on X by

$$f^y(x) = f(x, y), \qquad x \in X.$$

10.6 LEMMA. (a) *If E is a measurable subset of Z, then every section of E is measurable.*

(b) *If f is a measurable function on Z to $\bar{\boldsymbol{R}}$, then every section of f is measurable.*

PROOF. (a) If $E = A \times B$ and $x \in X$, then the x-section of E is B if $x \in A$, and is $\emptyset$ if $x \notin A$. Therefore, the rectangles are contained in the collection $\boldsymbol{E}$ of sets in $\boldsymbol{Z}$ having the property that each x-section is measurable. Since it is easily seen that $\boldsymbol{E}$ is a σ-algebra (see Exercise 10.I), it follows that $\boldsymbol{E} = \boldsymbol{Z}$.

(b) Let $x \in X$ and $\alpha \in \boldsymbol{R}$, then

$$\begin{aligned}\{y \in Y : f_x(y) > \alpha\} &= \{y \in Y : f(x, y) > \alpha\} \\ &= \{(x, y) \in X \times Y : f(x, y) > \alpha\}_x.\end{aligned}$$

If f is $\boldsymbol{Z}$-measurable, then f_x is $\boldsymbol{Y}$-measurable. Similarly, f^y is $\boldsymbol{X}$-measurable.

Q.E.D.

We interpolate an important result, which is often useful in measure and probability theory, and which will be used below. We recall (see Exercise 2.V) that a monotone class is a nonempty collection $\boldsymbol{M}$ of sets which contains the union of each increasing sequence in $\boldsymbol{M}$ and the intersection of each decreasing sequence in $\boldsymbol{M}$. It is easy (see Exercise 2.W) to show that if $\boldsymbol{A}$ is a nonempty collection of subsets of a set S, then the σ-algebra $\boldsymbol{S}$ generated by $\boldsymbol{A}$ contains the monotone class $\boldsymbol{M}$ generated by $\boldsymbol{A}$. We now show that if $\boldsymbol{A}$ is an algebra, then $\boldsymbol{S} = \boldsymbol{M}$.

10.7 MONOTONE CLASS LEMMA. *If $\boldsymbol{A}$ is an algebra of sets, then the σ-algebra $\boldsymbol{S}$ generated by $\boldsymbol{A}$ coincides with the monotone class $\boldsymbol{M}$ generated by $\boldsymbol{A}$.*

PROOF. We have remarked that $\boldsymbol{M} \subseteq \boldsymbol{S}$. To obtain the opposite inclusion it suffices to prove that $\boldsymbol{M}$ is an algebra.

If $E \in \boldsymbol{M}$, define $\boldsymbol{M}(E)$ to be the collection of $F \in \boldsymbol{M}$ such that $E \setminus F$, $E \cap F$, $F \setminus E$ all belong to $\boldsymbol{M}$. Evidently $\emptyset, E \in \boldsymbol{M}(E)$ and it is readily seen that $\boldsymbol{M}(E)$ is a monotone class. Moreover, $F \in \boldsymbol{M}(E)$ if and only if $E \in \boldsymbol{M}(F)$.

If E belongs to the algebra $\boldsymbol{A}$, then it is clear that $\boldsymbol{A} \subseteq \boldsymbol{M}(E)$. But since $\boldsymbol{M}$ is the smallest monotone class containing $\boldsymbol{A}$, we must have $\boldsymbol{M}(E) = \boldsymbol{M}$ for E in $\boldsymbol{A}$. Therefore, if $E \in \boldsymbol{A}$ and $F \in \boldsymbol{M}$, then $F \in \boldsymbol{M}(E)$. We infer that if $E \in \boldsymbol{A}$ and $F \in \boldsymbol{M}$, then $E \in \boldsymbol{M}(F)$ so that $\boldsymbol{A} \subseteq \boldsymbol{M}(F)$ for any $F \in \boldsymbol{M}$. Using the minimality of $\boldsymbol{M}$ once more we conclude that $\boldsymbol{M}(F) = \boldsymbol{M}$ for any $F \in \boldsymbol{M}$. Thus $\boldsymbol{M}$ is closed under intersections and relative complements. But since $X \in \boldsymbol{M}$ it is plain that $\boldsymbol{M}$ is an algebra; since it is a monotone class, it is indeed a σ-algebra. Q.E.D.

It follows from the Monotone Class Lemma that if a monotone class contains an algebra $\boldsymbol{A}$, then it contains the σ-algebra generated by $\boldsymbol{A}$.

10.8 LEMMA. *Let $(X, \boldsymbol{X}, \mu)$ and $(Y, \boldsymbol{Y}, \nu)$ be σ-finite measure spaces. If $E \in \boldsymbol{Z} = \boldsymbol{X} \times \boldsymbol{Y}$, then the functions defined by*

$$f(x) = \nu(E_x), \quad g(y) = \mu(E^y) \tag{10.2}$$

are measurable, and

$$\int_X f \, d\mu = \pi(E) = \int_Y g \, d\nu. \tag{10.3}$$

PROOF. First we shall suppose that the measure spaces are finite and let $\boldsymbol{M}$ be the collection of all $E \in \boldsymbol{Z}$ for which the above assertion is true. We shall show that $\boldsymbol{M} = \boldsymbol{Z}$ by demonstrating that $\boldsymbol{M}$ is a monotone class containing the algebra $\boldsymbol{Z}_0$. In fact, if $E = A \times B$ with $A \in \boldsymbol{X}$ and $B \in \boldsymbol{Y}$, then

$$f(x) = \chi_A(x)\,\nu(B), \qquad g(y) = \chi_B(y)\,\mu(A),$$
$$\int_X f \, d\mu = \mu(A)\,\nu(B) = \int_Y g \, d\nu.$$

Since an arbitrary element of $\boldsymbol{Z}_0$ can be written as a finite disjoint union of rectangles, it follows that $\boldsymbol{Z}_0 \subseteq \boldsymbol{M}$.

We now show that $\boldsymbol{M}$ is a monotone class. Indeed, let (E_n) be a monotone increasing sequence in $\boldsymbol{M}$ with union E. Therefore

$$f_n(x) = \nu\big((E_n)_x\big), \qquad g_n(y) = \mu\big((E_n)^y\big)$$

are measurable and

$$\int_X f_n \, d\mu = \pi(E_n) = \int_Y g_n \, d\nu.$$

It is clear that the monotone increasing sequences (f_n) and (g_n) converge to the functions f and g defined by

$$f(x) = \nu(E_x), \qquad g(y) = \mu(E^y).$$

If we apply the fact that π is a measure and the Monotone Convergence Theorem, we obtain

$$\int_X f \, d\mu = \pi(E) = \int_Y g \, d\nu,$$

so that $E \in \boldsymbol{M}$. Since π is finite measure, it can be proved in the same way that if (F_n) is a monotone decreasing sequence in $\boldsymbol{M}$, then $F = \bigcap F_n$ belongs to $\boldsymbol{M}$. Therefore $\boldsymbol{M}$ is a monotone class, and it follows from the Monotone Class Lemma that $\boldsymbol{M} = \boldsymbol{Z}$.

If the measure spaces are σ-finite, let Z be the increasing union of a sequence of rectangles (Z_n) with $\pi(Z_n) < +\infty$ and apply the previous argument and the Monotone Convergence Theorem to the sequence $(E \cap Z_n)$. Q.E.D.

10.9 TONELLI'S THEOREM. *Let $(X, \boldsymbol{X}, \mu)$ and $(Y, \boldsymbol{Y}, \nu)$ be σ-finite measure spaces and let F be a nonnegative measurable function on $Z = X \times Y$ to $\bar{\boldsymbol{R}}$. Then the functions defined on X and Y by*

$$f(x) = \int_Y F_x \, d\nu, \qquad g(y) = \int_X F^y \, d\mu, \tag{10.4}$$

are measurable and

$$\int_X f \, d\mu = \int_Z F \, d\pi = \int_Y g \, d\nu. \tag{10.5}$$

In other symbols,

$$\int_X \left(\int_Y F \, d\nu \right) d\mu = \int_Z F \, d\pi = \int_Y \left(\int_X F \, d\mu \right) d\nu. \tag{10.6}$$

PROOF. If F is the characteristic function of a set in $\boldsymbol{Z}$, the assertion follows from the Lemma 10.8. By linearity, the present theorem holds for a measurable simple function. If F is an arbitrary nonnegative measurable function on Z to $\bar{\boldsymbol{R}}$, Lemma 2.11 implies that there is a sequence (Φ_n) of nonnegative measurable simple functions which converges in a monotone increasing fashion on Z to F. If φ_n and ψ_n are defined by

$$(10.7) \qquad \varphi_n(x) = \int_Y (\Phi_n)_x \, d\nu, \qquad \psi_n(y) = \int_X (\Phi_n)^y \, d\mu,$$

then φ_n and ψ_n are measurable and monotone in n. By the Monotone Convergence Theorem, (φ_n) converges on X to f and (ψ_n) converges on Y to g. Another application of the Monotone Convergence Theorem implies that

$$\begin{aligned} \int_X f \, d\mu &= \lim \int_X \varphi_n \, d\mu = \lim \int_Z \Phi_n \, d\pi \\ &= \lim \int_Y \psi_n \, d\nu = \int_Y g \, d\nu. \end{aligned}$$

The same theorem also shows that

$$\int_Z F \, d\pi = \lim \int_Z \Phi_n \, d\pi,$$

from which (10.5) follows. Q.E.D.

It will be seen in the exercises that Tonelli's Theorem may fail if we drop the hypothesis that F is nonnegative, or if we drop the hypothesis that the measures μ, ν are σ-finite.

Tonelli's Theorem deals with a nonnegative function on Z and affirms the equality of the integral over Z and the two iterated integrals whether these integrals are finite or equal $+\infty$. The final result considers the case where the function is allowed to take both positive and negative values, but is assumed to be integrable.

10.10 Fubini's Theorem. *Let $(X, \boldsymbol{X}, \mu)$ and $(Y, \boldsymbol{Y}, \nu)$ be σ-finite spaces and let the measure π on $\boldsymbol{Z} = \boldsymbol{X} \times \boldsymbol{Y}$ be the product of μ and ν. If the function F on $Z = X \times Y$ to $\boldsymbol{R}$ is integrable with respect to π, then the extended real-valued functions defined almost everywhere by*

$$(10.8) \qquad f(x) = \int_Y F_x \, d\nu, \qquad g(y) = \int_X F^y \, d\mu$$

have finite integrals and

$$(10.9) \qquad \int_X f\,d\mu = \int_Z F\,d\pi = \int_Y g\,d\nu.$$

In other symbols,

$$(10.10) \qquad \int_X \left[\int_Y F\,d\nu\right] d\mu = \int_Z F\,d\pi = \int_Y \left[\int_X F\,d\mu\right] d\nu.$$

PROOF. Since F is integrable with respect to π, its positive and negative parts F^+ and F^- are integrable. Apply Tonelli's Theorem to F^+ and F^- to deduce that the corresponding f^+ and f^- have finite integrals with respect to μ. Hence f^+ and f^- are finite-valued μ-almost everywhere, so their difference f is defined μ-almost everywhere and the first part of (10.9) is clear. The second part is similar.

Q.E.D.

Since we have chosen in Chapter 5 to restrict the use of the word "integrable" to real-valued functions, we cannot conclude that the functions f, g defined in (10.8) are integrable. However, they are almost everywhere equal to integrable functions.

It will be seen in an exercise that Fubini's Theorem may fail if the hypothesis that F is integrable is dropped.

EXERCISES

10.A. Let $A \subseteq X$ and $B \subseteq Y$. If A or B is empty, then $A \times B = \emptyset$. Conversely, if $A \times B = \emptyset$, then either $A = \emptyset$ or $B = \emptyset$.

10.B. Let $A_j \subseteq X$ and $B_j \subseteq Y$, $j = 1, 2$. If $A_1 \times B_1 = A_2 \times B_2$, then $A_1 = A_2$ and $B_1 = B_2$.

10.C. Let $A_j \subseteq X$ and $B_j \subseteq Y$, $j = 1, 2$. Then

$$(A_1 \times B_1) \cup (A_2 \times B_2) = [(A_1 \setminus A_2) \times B_1] \cup [(A_1 \cap A_2) \times (B_1 \cup B_2)] \cup [(A_2 \setminus A_1) \times B_2],$$

and the sets on the right side are mutually disjoint.

10.D. Let $(X, \boldsymbol{X})$ and $(Y, \boldsymbol{Y})$ be measurable spaces. If $A_j \in \boldsymbol{X}$ and $B_j \in \boldsymbol{Y}$ for $j = 1, \ldots, m$, then the set

$$\bigcup_{j=1}^{n} (A_j \times B_j)$$

can be written as the *disjoint* union of a finite number of rectangles in Z.

10.E. Let $A_j \subseteq X$ and $B_j \subseteq Y, j = 1, 2$. Then

$$(A_1 \times B_1) \setminus (A_2 \times B_2) = [(A_1 \cap A_2) \times (B_1 \setminus B_2)] \cup [(A_1 \setminus A_2) \times B_1]$$

$$(A_1 \times B_1) \cap (A_2 \times B_2) = (A_1 \cap A_2) \times (B_1 \cap B_2).$$

10.F. If $(\boldsymbol{R}, \boldsymbol{B})$ denotes the measurable space consisting of real numbers together with the Borel sets, show that every open subset of $\boldsymbol{R} \times \boldsymbol{R}$ belongs to $\boldsymbol{B} \times \boldsymbol{B}$. In fact, this σ-algebra is the σ-algebra generated by the open subsets of $\boldsymbol{R} \times \boldsymbol{R}$. (In other words, $\boldsymbol{B} \times \boldsymbol{B}$ is the Borel algebra of $\boldsymbol{R} \times \boldsymbol{R}$.)

10.G. Let f and g be real-valued functions on X and Y, respectively; suppose that f is $\boldsymbol{X}$-measurable and that g is $\boldsymbol{Y}$-measurable. If h is defined for (x, y) in $X \times Y$ by $h(x, y) = f(x)\,g(y)$, show that h is $\boldsymbol{X} \times \boldsymbol{Y}$-measurable.

10.H. If E is a subset of $\boldsymbol{R}$, let $\gamma(E) = \{(x, y) \in \boldsymbol{R} \times \boldsymbol{R} : x - y \in E\}$. If $E \in \boldsymbol{B}$, show that $\gamma(E) \in \boldsymbol{B} \times \boldsymbol{B}$. Use this to prove that if f is a Borel measurable function on $\boldsymbol{R}$ to $\boldsymbol{R}$, then the function F defined by $F(x, y) = f(x - y)$ is measurable with respect to $\boldsymbol{B} \times \boldsymbol{B}$.

10.I. Let E and F be subsets of $Z = X \times Y$, and let $x \in X$. Show that $(E \setminus F)_x = E_x \setminus F_x$. If (E_α) are subsets of Z, then

$$(\bigcup E_\alpha)_x = \bigcup (E_\alpha)_x.$$

10.J. Let $(X, \boldsymbol{X}, \mu)$ be the measure space on the natural numbers $X = \boldsymbol{N}$ with the counting measure defined on all subsets of $X = \boldsymbol{N}$. Let $(Y, \boldsymbol{Y}, \nu)$ be an arbitrary measure space. Show that a set E in $Z = X \times Y$ belongs to $\boldsymbol{Z} = \boldsymbol{X} \times \boldsymbol{Y}$ if and only if each section E_n of E belongs to $\boldsymbol{Y}$. In this case there is a unique product measure π, and

$$\pi(E) = \sum_{n=1}^{\infty} \nu(E_n), \qquad E \in \boldsymbol{Z}.$$

A function f on $Z = X \times Y$ to $\boldsymbol{R}$ is measurable if and only if each section f_n is $\boldsymbol{Y}$-measurable. Moreover, f is integrable with respect to π if and only if the series

$$\sum_{n=1}^{\infty} \int_Y |f_n| \, d\nu$$

is convergent, in which case

$$\int_Z f \, d\pi = \sum_{n=1}^{\infty} \left[\int_Y f_n \, d\nu \right] = \int_Y \left[\sum_{n=1}^{\infty} f_n \right] d\nu .$$

10.K. Let X and Y be the unit interval $[0, 1]$ and let $\boldsymbol{X}$ and $\boldsymbol{Y}$ be the Borel subsets of $[0, 1]$. Let μ be Lebesgue measure on $\boldsymbol{X}$ and let ν be the counting measure on $\boldsymbol{Y}$. If $D = \{(x, y) : x = y\}$, show that D is a measurable subset of $Z = X \times Y$, but that

$$\int \nu(D_x) \, d\mu(x) \neq \int \mu(D^y) \, d\nu(y).$$

Hence Lemma 10.8 may fail unless both of the factors are required to be σ-finite.

10.L. If F is the characteristic function of the set D in the Exercise 10.K, show that Tonelli's Theorem may fail unless both of the factors are required to be σ-finite.

10.M. Show that the example considered in Exercise 10.J demonstrates that Tonelli's Theorem holds for arbitrary $(Y, \boldsymbol{Y}, \nu)$ when $(X, \boldsymbol{X}, \mu)$ is the set $\boldsymbol{N}$ of natural numbers with the counting measure on arbitrary subsets of $\boldsymbol{N}$.

10.N. If $a_{mn} \geqslant 0$ for $m, n \in \boldsymbol{N}$, then

$$\sum_{m=1}^{\infty} \sum_{n=1}^{\infty} a_{mn} = \sum_{n=1}^{\infty} \sum_{m=1}^{\infty} a_{mn} \quad (\leqslant +\infty).$$

10.O. Let a_{mn} be defined for $m, n \in \boldsymbol{N}$ by requiring that $a_{nn} = +1$, $a_{n,n+1} = -1$, and $a_{mn} = 0$ if $m \neq n$ or $m \neq n + 1$. Show that

$$\sum_{m=1}^{\infty} \sum_{n=1}^{\infty} a_{mn} = 0, \qquad \sum_{n=1}^{\infty} \sum_{m=1}^{\infty} a_{mn} = 1,$$

so the hypothesis of integrability in Fubini's Theorem cannot be dropped.

10.P. Let f be integrable on $(X, \boldsymbol{X}, \mu)$, let g be integrable on $(Y, \boldsymbol{Y}, \nu)$, and define h on Z by $h(x, y) = f(x)\, g(y)$. If π is a product of μ and ν, show that h is π-integrable and

$$\int_Z h \, d\pi = \left[\int_X f \, d\mu\right]\left[\int_Y g \, d\nu\right].$$

10.Q. Suppose that $(X, \boldsymbol{X}, \mu)$ and $(Y, \boldsymbol{Y}, \nu)$ are σ-finite, and let E, F belong to $\boldsymbol{X} \times \boldsymbol{Y}$. If $\nu(E_x) = \nu(F_x)$ for all $x \in X$, then $\pi(E) = \pi(F)$.

10.R. Let f and g be Lebesgue integrable functions on $(\boldsymbol{R}, \boldsymbol{B})$ to $\boldsymbol{R}$. From Exercise 10.H it follows that the function mapping (x, y) into $f(x - y)\, g(y)$ is measurable with respect to $\boldsymbol{B} \times \boldsymbol{B}$. If λ denotes Lebesgue measure on $\boldsymbol{B}$, use Tonelli's Theorem and the fact that

$$\int_{\boldsymbol{R}} |f(x - y)| \, d\lambda(x) = \int_{\boldsymbol{R}} |f(x)| \, d\lambda(x)$$

to show that the function h defined for $x \in \boldsymbol{R}$ by

$$h(x) = \int_{\boldsymbol{R}} f(x - y)\, g(y) \, d\lambda(y)$$

is finite almost everywhere. Moreover,

$$\int |h| \, d\lambda \leqslant \left[\int |f| \, d\lambda\right]\left[\int |g| \, d\lambda\right].$$

The function h defined above is called the **convolution** of f and g and is usually denoted by $f * g$.

10.S. Let $X = \boldsymbol{R}$, $\boldsymbol{X}$ be the σ-algebra of all subsets of $\boldsymbol{R}$ and let μ be defined by $\mu(A) = 0$ if A is countable, and $\mu(A) = +\infty$ if A is uncountable. We shall construct distinct products of μ with itself.

(a) If $E \in \boldsymbol{Z} = \boldsymbol{X} \times \boldsymbol{X}$, define $\pi(E) = 0$ in case E can be written as the union $E = G \cup H$ of two sets in $\boldsymbol{Z}$ such that the x-projection of G is countable and the y-projection of H is countable. Otherwise, define $\pi(E) = +\infty$. It is evident that π is a measure on $\boldsymbol{Z}$. If $\pi(E) = 0$, then E is contained in the union of a countable set of lines in the plane. If $A, B \in \boldsymbol{X}$, show that $\pi(A \times B) = \mu(A)\, \mu(B)$. Hence π is a product of μ with itself.

(b) If $E \in \boldsymbol{Z}$, define $\rho(E) = 0$ in case E can be written as the union $E = G \cup H \cup K$ of three sets in $\boldsymbol{Z}$ such that the x-projection of G is countable, the y-projection of H is countable, and the projection of K on the line with equation $y = x$ is countable. Otherwise, define $\rho(E) = +\infty$. Now ρ is a measure of $\boldsymbol{Z}$, and if $\rho(E) = 0$, then E is contained in the union of a countable set of lines. Show that $\rho(A \times B) = \mu(A)\,\mu(B)$ for all $A, B \in \boldsymbol{X}$; hence ρ is a product of μ with itself.

(c) Let $E = \{(x, y) : x + y = 0\}$; show that $E \in \boldsymbol{Z}$. However, $\rho(E) = 0$, whereas $\pi(E) = +\infty$.

References

1. Bartle, R. G., *The Elements of Real Analysis*, John Wiley and Sons, New York, 1964.
2. Bartle, R. G., "A General Bilinear Vector Integral," *Studia Mathematica*, Vol. **15** (1956).
3. Berberian, S. K., *Measure and Integration*, Macmillan, New York, 1965.
4. Bourbaki, N., *Éléments de Mathématique*, Livre VI, *Intégration*, Hermann et Cie, Act. Sci. et Ind., 1175, 1244, 1281, 1306, Paris, 1952, 1956, 1959, 1963.
5. Dunford, N., and J. T. Schwartz, *Linear Operators*, Part I, Interscience, New York, 1958.
6. Graves, L. M., *Theory of Functions of Real Variables*, Second edition, McGraw-Hill, New York, 1956.
7. Halmos, P. R., *Measure Theory*, D. Van Nostrand, New York, 1950.
8. Loomis, L. H., *An Introduction to Abstract Harmonic Analysis*, D. Van Nostrand, New York, 1953.
9. McShane, E. J., *Integration*, Princeton University Press, New Jersey, 1944.
10. Munroe, M. E., *Introduction to Measure and Integration*, Addison-Wesley, Cambridge, Mass., 1953.
11. Naimark, M. A., *Normed Rings*, (Translation from Russian) Noordhoff, Groningen, 1959.
12. Royden, H. L., *Real Analysis*, Macmillan, New York, 1963.
13. Saks, S., *Theory of the Integral*, Second edition, Monografje Matematyczne, Vol. 7, Warsaw, 1937. Reprinted by Hafner, New York.
14. Schaefer, H. H., "A Brief Introduction to the Lebesgue–Stieltjes

Integral," Vol. **3** (1965), in *Studies in Real and Complex Analysis*, I. I. Hirschman, Jr., editor. Published by the Mathematical Association of America.

15. Stone, M. H., "Notes on Integration," *Proceedings Nat. Acad. Sci. U.S.A.*, Vols. **34** (1948) and **35** (1949).
16. Taylor, A. E., *General Theory of Functions and Integration*, Blaisdell, New York, 1965.
17. Titchmarsh, E. C., *The Theory of Functions*, Second edition, Oxford University Press, London, 1939.
18. Zaanen, A. C., *An Introduction to the Theory of Integration*, Interscience, New York, 1958.

Index